Root Crops in Eastern Africa

Proceedings of a workshop held in Kigali, Rwanda, 23–27 November 1980

Cosponsored by Gouvernement de la République rwandaise, the International Institute of Tropical Agriculture, and the International Development Research Centre.

Résumé

Cette brochure traite principalement des deux tubercules alimentaires les plus importants en Afrique orientale, soit le manioc et la patate douce. Quelques communications portent sur la pomme de terre, l'igname, le taro et l'« enset » dont la consommation est considérable dans plusieurs pays de la région. Le rendement de ces cultures est limité par de nombreux facteurs. Aussi, la recherche effectuée dans le cadre de programmes agronomiques nationaux et internationaux est-elle orientée vers la correction de cette situation en Afrique. Les difficultés rencontrées en cours de travaux et les progrès réalisés sont décrits par des représentants et des consultants de l'Institut international d'agriculture tropicale d'Ibadan (Nigéria) et d'autres pays tel que le Cameroun, le Kenya, l'Ouganda, le Malawi, le Zimbabwe, l'Éthiopie, le Burundi, le Zaïre et le Swaziland.

Resumen

Esta publicación se enfoca en la mandioca y el camote — los cultivos de tuberosas más importantes del Africa oriental. Los trabajos tratan también del *Solanum tuberosum*, *Dioscorea* spp., *Colocasia* sp., *Xanthosoma* sp., y *Enset* sp., que son todos cultivos importantes a los países de esta región. La producción de cada uno es restringida por serios constreñimientos, y el alivio de éstos es el objetivo de varias investigaciones llevadas a cabo por los programas agrícolas nacionales e internacionales en el Africa. El progreso hacia y los problemas encontrados en llegar a este fin son delineados por especialistas representando al Instituto Internacional de Agricultura Tropical en Ibadan, Nigeria, y a los países de Camerún, Kenia, Uganda, Malawi, Zimbabwe, Etiopia, Burundi, Zaire, y Swazilandia.

Contents

Country reports

Foreword

The Eastern Africa root-crop workshop was concerned with cassava, sweet potato, yams, and cocoyams, with particular emphasis on the first two crops because of their relative importance in Eastern Africa. None of these crops is indigenous to the area. Cassava was probably introduced from Latin America to the lowland coastal region of Eastern Africa by sea during the 18th century and to the highland regions of lakes Edward, Kwu, Victoria, Tanganyika, Malawi, and Bangwudu through western Zaire at about the same time or a little earlier. The similarity in the use of the crop within coastal and highland regions and the presence of bacterial blight only in the highland region suggest that it was introduced into Eastern Africa from two directions. When and how sweet potato was introduced into Eastern Africa are uncertain. It may, however, have been grown in the region before the arrival of Europeans. Although white yam (*Dioscorea rotundata*) and yellow yam (*D. cayenensis*) were domesticated in West Africa, they must have been introduced into Eastern Africa comparatively recently. The Asian water yam (*D. alata*) is more widely grown in Eastern Africa than is the African yam. The old cocoyam (*Colocasia* spp.), which was domesticated in Asia, was probably introduced into Eastern Africa during the 16th century through Egypt. The new cocoyam (*Xanthosoma* spp.), which was domesticated in Central America, was probably introduced much later.

Cassava and sweet potato were quickly and widely adapted; they became important staple-food crops in the traditional farming systems in different agroecological areas. They have the ability to adapt to diverse environmental conditions and farming systems. They can produce high yields in adverse conditions, can be grown with limited inputs, and are drought tolerant, surviving even 2–4 months of dry weather. When conditions are favourable, they can recover from damage caused by insect pests and diseases. Since their introduction, they have played a vital role in alleviating famine conditions in the region by providing a sustained food supply when other crops fail. The leaves of both crops are used as a vegetable and are a major source of protein, vitamins, and minerals; the storage roots are used by the population as the major source of carbohydrate. One often forgets the merits of these crops when conditions are favourable and other crops flourish. However, as the recent drought in the region showed, these crops are like lifeblood and deserve special attention.

Active research and development are now taking place to use root crops for energy in countries that do not produce fuel, projects already being under way in Brazil, Japan, and Australia. Countries in Eastern Africa will also consider these projects in the future.

However, at present, yields of the root crops in the region are far too low to meet the demand and do not reflect the potential. Researchers have demonstrated that yields can be increased substantially through improved cultural practices and improved varieties.

The importance of these root crops is, slowly but steadily, being realized by government officials in many countries in Eastern Africa, although the opportunities to exchange research findings and views among researchers and policymakers are few. This Eastern Africa root-crop workshop was, therefore, held so that researchers and policymakers could:

- Discuss and identify the constraints to root-crop production in the region;
- Establish means to work together internationally and nationally toward a solution of the production constraints; and
- Consider future development and research in root crops in the region.

The need for stock-taking and exchange among researchers so that they can move forward toward common goals has been recognized and increasingly supported by funding agencies. The International Development Research Centre (Canada) and the Ford Foundation (USA), therefore, agreed to provide funds for the Eastern Africa root-crop workshop, the proceedings of which constitute this publication.

S.K. Hahn
International Institute of Tropical Agriculture

and

Andrew D.R. Ker
Agriculture, Food and Nutrition Sciences Division
IDRC

Participants

B.D.A. Beck Brian Beck (Pvt) Ltd, Box 86, Zomba, Malawi

Moshi Bititaweho Université nationale du Rwanda (UNR), B.P. 117, Butare, Rwanda

Innocent Butare Institut des sciences agronomiques du Rwanda (ISAR), B.P. 138, Buture, Rwanda

A.J. Carpenter Food and Agriculture Organization of the United Nations (FAO), P.O. Box 159, Zanzibar, Tanzania

D. Cimpaye Institut des sciences agronomiques du Burundi (ISABU), B.P. 795, Bujumbura, Burundi

G.H. de Bruijn Department of Crop Science, University of Nairobi, P.O. Box 30197, Nairobi, Kenya

Gilbert Delepierre Programme belge de recherches agronomiques au Rwanda (PBRA), B.P. 1051, Kigali, Rwanda

A. Eid Food and Agriculture Organization of the United Nations (FAO), B.P. 445, Kigali, Rwanda

Thaddée Gatarasi Institut des sciences agronomiques du Rwanda (ISAR), B.P. 138, Butare, Rwanda

W. Godfrey-Sam-Aggrey Department of Crop Production, University College of Swaziland, P.O. Luyengo, Swaziland

Michael Graham Communications Division, International Development Research Centre (IDRC), Tanglin P.O. Box 101, Singapore 9124, Republic of Singapore

E.J. Guthrie Overseas Development Administration, Crop Virology Research Project, Kenya Agricultural Research Institute, P.O. Box 30148, Nairobi, Kenya

S. Habyalimana Bureau national d'études des projets (BNEP), B.P. 1337, Kigali, Rwanda

S.K. Hahn International Institute of Tropical Agriculture (IITA), P.M.B. 5320, Ibadan, Nigeria

Faustin Iyamuremye Institut des sciences agronomiques du Rwanda (ISAR), B.P. 138, Butare, Rwanda

M.J.J. Janssens Institut des sciences agronomiques du Rwanda (ISAR), B.P. 138, Butare, Rwanda

Emmanuel Kayitare Institut des sciences agronomiques du Rwanda (ISAR), B.P. 138, Butare, Rwanda

Andrew D.R. Ker Agriculture, Food and Nutrition Sciences Division, International Development Research Centre (IDRC), P.O. Box 8500, Ottawa, Canada K1G 3H9

K.M. Lema International Institute of Tropical Agriculture (IITA), P.M.B. 5320, Ibadan, Nigeria

K. Leuschner International Institute of Tropical Agriculture (IITA), P.M.B. 5320, Ibadan, Nigeria

N.B. Lutaladio Programme national manioc (PRONAM), B.P. 11635, Kinshasa, Zaire

S.N. Lyonga Cameroon National Root Crops Improvement Program (CNRCIP), Institut de la recherche agronomique (IRA) Ekona Centre, P.M.B. 25, Nuea, South West Province, Cameroon

Apollinaire Mpabanzi Institut des sciences agronomiques du Rwanda (ISAR), B.P. 138, Butare, Rwanda

J. Mulindangabo Institut des sciences agronomiques du Rwanda (ISAR), B.P. 121, Kigali, Rwanda

Thaddée Musabyimana CIP/Programme national de l'amélioration de la pomme de terre (PNAP), B.P. 73, Ruhengeri, Rwanda

Georges Ndamage Institut des sciences agronomiques du Rwanda (ISAR), B.P. 121, Kigali, Rwanda

Innocent Ndindabahizi Université national du Rwanda (UNR), B.P. 117, Butare, Rwanda

Henri Neel Ambassade de Belgique, B.P. 81, Kigali, Rwanda

Dismas Nsabimana Ministère de l'agriculture et de l'élevage (MINAGRI), B.P. 621, Kigali, Rwanda

Pierre Nyabyenda Institut des sciences agronomiques du Rwanda (ISAR), B.P. 138, Butare, Rwanda

Z.M. Nyiira Kawanda Research Station, Ministry of Agriculture, P.O. Box 7065, Kampala, Uganda

Frédéric Nzamurambaho Ministère de l'agriculture et de l'élevage (MINAGRI), B.P. 621, Kigali, Rwanda

A. Nzindukiyimana Bulletin agricole du Rwanda, B.P. 104, Kigali, Rwanda

Anastase Nzirasanaho Ministère de l'agriculture et de l'élevage (MINAGRI), B.P. 621, Kigali, Rwanda

D.M. Onyango National Agricultural Laboratories (NAL), P.O. Box 14733, Nairobi, Kenya

René Paquay Service des semences sélectionnées (SSS), B.P. 538, Kigali, Rwanda

H.J. Pfeiffer Cameroon National Root Crops Improvement Program (CNRCIP), Institut de la recherche agronomique (IRA), B.P. 2067, Yaounde, Cameroon

A.G. Rowe Department of Research and Specialist Services, P.O. Box 8100, Causeway, Harare, Zimbabwe

Venant Rutunga Institut des sciences agronomiques du Rwanda (ISAR), B.P. 138, Butare, Rwanda

R.F.Nembozanga Sauti Ministry of Agriculture and Natural Resources, Lunyangwa Agricultural Research Station, P.O. Box 59, Mzuzu, Malawi

Claudien Sekanabanga Service des semences sélectionnées (SSS), B.P. 538, Kigali, Rwanda

T.P. Singh Programme national manioc (PRONAM), B.P. 11635, Kinshasa, Zaire

Pierre Tegera CIP/Programme national de l'amélioration de la pomme de terre (PNAP), B.P. 73, Ruhengeri, Rwanda

Terefe Belehu Institute of Agricultural Research (IAR), P.O. Box 2003, Addis Ababa, Ethiopia

E.R. Terry International Institute of Tropical Agriculture (IITA), P.M.B. 5320, Ibadan, Nigeria

P. Vander Zaag CIP/Programme national de l'amélioration de la pomme de terre (PNAP), Institut des sciences agronomiques du Rwanda, B.P. 73, Ruhengeri, Rwanda

Christian Verstraete Food and Agriculture Organization of the United Nations (FAO), B.P. 445, Kigali, Rwanda

Discussion summary

The major root crops in the region are: cassava (*Manihot esculenta*), sweet potato (*Ipomoea batatas*), cocoyam (*Xanthosoma* spp. and *Colocasia* spp.), and yams (*Dioscorea* spp.); potatoes (*Solanum tuberosum*) and enset (*Enset ventricosa*) are considered to be of minor importance.

Major production constraints for all root crops include: environmental extremes, such as temperature, moisture, and photoperiod; wide ranges in soil fertility; a multiplicity of pest and disease agents; wide use of unimproved local varieties; absence of viable cropping systems that would ensure maximum productivity of root crops; lack of adequate dissemination of available technology; and a shortage of properly trained personnel.

To overcome these constraints, there is a need for: an inventory of local cultivars for studies on their inherent genetic potential; the utilization of already-developed cultivars that have high-yield potentials and high levels of disease and pest resistance; and the use of available information on cultural practices, such as crop hygiene.

The areas of research that were assigned high priority by the participants at the Eastern Africa root-crops workshop were: defining the relevant environmental, socioeconomic, and agronomic constraints to optimum productivity; developing viable cropping systems based on well-defined criteria; establishing economic models to determine the investment potential of different cropping systems; developing and testing improved cultivars with high-yield potential, pest and disease resistance, and good consumer acceptability and storability; designing storage systems appropriate for local conditions; and formulating a wider range of processing techniques for more efficient utilization of present production.

To improve personnel capability for future root-crops research, one must exploit the comparative advantage of available research institutions, thus making the most efficient use of available resources and avoiding duplication. Training programs should be tailored to meet locally defined priorities, and major emphasis should be placed on developing strong national research programs, which can later form the basis for regional cooperation. Donor agencies should concentrate their efforts primarily on encouraging strong national programs and, thereafter, assisting with the development of regional programs and fostering links with relevant international agricultural research centres.

There is a need to foster improved communication between researchers and also between researchers and policymakers to ensure that funds are allocated to areas that require attention urgently.

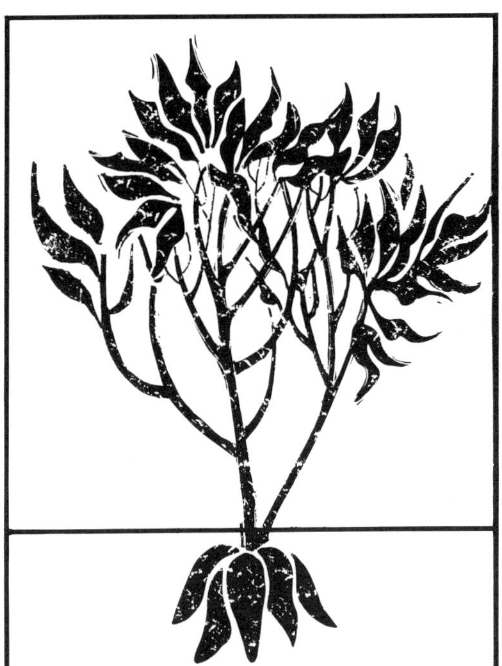

Breeding

Cassava bacterial blight — a focus of breeding programs in Africa — at work on a susceptible cultivar.

Historical perspectives of cassava breeding in Africa

B.D.A. Beck[1]

The early records of the introduction of cassava to East and West Africa are reviewed, and the three major cassava-breeding programs that involved the use of interspecific hybridization to obtain resistance to cassava mosaic disease (CMD) are outlined. My earlier work to produce the interspecies hybrid clone 58308 is described; this clone has shown stable resistance to CMD for 22 years under continuous, high-inoculation pressure. Researchers at the International Institute of Tropical Agriculture have shown that this resistance is recessive, is polygenic, and is inherited in a largely additive manner. They have also shown that 58308 provides polygenic resistance to cassava bacterial blight (CBB) and that this resistance is also recessive and is inherited additively. The resistances to CMD and CBB are strongly linked, and this link may occur in the chromosome complement derived from the *Manihot glaziovii* parent. Clone 58308 has also been the principal parent in the production of lines with low cyanide in the roots. Resistance to both diseases and the presence of low cyanide have been incorporated into a large number of high-yielding populations at IITA, through a half-sib selection procedure with relatively large families. The future development and the distribution of new disease-resistant, low-cyanide cultivars for the smallholder and for large-scale producers as an industrial feedstock from these populations are discussed.

Revue de l'historique de l'introduction du manioc en Afrique orientale et occidentale et aperçu des trois principaux programmes de sélection génétique du manioc ayant recours à l'hybridation inter spécifique pour obtenir une espèce résistante à la mosaïque. Description des expériences de production du clone hybride inter spécifique 58308 ; ce clone a fait preuve d'une résistance stable à la mosaïque pendant vingt-deux ans de forte inoculation continuelle. Les chercheurs de l'Institut international pour l'agriculture tropicale ont montré que cette résistance est récessive, polygénique et héritée selon un mode hautement additif. Ils ont aussi démontré que le 58308 a une résistance polygénique à la brûlure bactérienne également récessive et héritée de façon additive. Les résistances à la mosaïque et à la brûlure bactérienne sont fortement liées, et ce lien peut se retrouver dans le complément de chromosome dérivé du parent *Manihot glaziovii*. Le clone 58308 est aussi la souche principale pour la production de lignées dont les racines ont peu de cyanure. La résistance aux deux maladies et le faible taux de cyanure ont été introduits à l'IITA dans de nombreuses populations à rendement élevé, par un processus de sélection de demi-frères au sein d'assez grandes familles. On discute des améliorations futures des nouveaux cultivars résistants aux maladies et à faible taux de cyanure, et de leur distribution comme fourrage industriel aux petits exploitants et aux grands producteurs.

Cassava (*Manihot esculenta*) originated in South America, and Vavilov (1949) suggested that cultivated strains spread from a Brazilian–Paraguayan centre of origin. Cassava was widely grown in Brazil in the 15th century, where Pierre Matyr first described the use of cassava roots in bread in 1494 (Pynaert 1928).

Cassava meal was used by the Portuguese to provision their slave ships (Hawkins 1593) and was brought to the West Coast of Africa by slave traders who planted the crop at their factories to feed slaves in the barracoons. Lacourbe (1635) recorded cassava growing at Bissau, and Dapper (1668) reported the crop at Fourcados in the Niger Delta. Cassava was widely grown by farmers in Owerri by 1700, but "nowhere else was it a crop of more than

[1] Brian Beck (Pvt) Ltd, Box 86, Zomba, Malawi.

local importance prior to the nineteenth century" (Jones 1957).

Cassava may have been introduced to Central Africa west of Lake Tanganyika along traditional overland trade routes from the Congo basin. The Portuguese brought manioc to the Island of Reunion from Brazil in 1739 and probably from there to Madagascar (IFAN 1951). Stuhlman (1909) quoted Strandes who stated that the first authenticated introduction to Mozambique was in 1750. Lacerda (1798) recorded buying cassava flour in Cazembe. Jones (1959, p. 84), discussing the reports of early travelers in East and Central Africa, suggested that, although there are many records of the occurrence of cassava in East Africa by 1850, it was at that time an unimportant crop, save on the coast and around Lake Tanganyika, to which it had been brought from the west.

The original importations of cassava to Africa were probably quite small and were obtained by the Portuguese from the Tupinamba Indians on the coast of Brazil who brought the crop from the high forest early in the 15th century (Jones 1959, p. 31). These cultivars were likely to have been adapted to the West African forest regions where slash-and-burn agriculture was widely established.

The fertility of forest soil cleared for cultivation falls rapidly during arable cropping. The ability of cassava, a highly efficient utilizer of soil nutrients, to grow on soils exhausted by previous cropping led to its cultivation as the last crop of the rotation before the land was returned to forest. A well-established cassava crop competes well with weeds, freeing the farmer to work a newly cleared area, returning to the cassava garden to harvest small groups of plants as they are required. Cassava was left in the field for up to 30 months with the minimum of attention. During this period, it flowered and produced seed, and the seedlings had time and opportunity to grow in the areas recently cleared by harvests. When the farmer returned to the cassava garden to gather planting material for the next cycle of cultivation, he or she probably used only some of the seedling material, selecting the outstanding types.

Although the original importations were small, they were highly heterozygous; the system of agriculture in which they were used lent itself to natural outcrossing and seedling propagation with the development and selection of new types. Thus, when I commenced a cassava-improvement program in Nigeria in 1954, I was able to collect more than 450 morphologically distinct local cultivars from farmers' fields in the country. A similar range of diversity has been found by other workers in Africa. However, the yield of most of the African cultivars was low, less than 20 t/ha, and all the local selections have proved susceptible to cassava mosaic disease (CMD).

CMD is an African virus. It has not been reported from Latin America, where the crop originated. CMD was first described in East Africa in 1885 (Lefevre 1935) and first recorded in West Africa, in Nigeria, by Jones in 1926 (Golding 1936). Golding, writing in 1936, stated that mosaic had been reported in "the Belgian Congo, Ubangin-Shai, Moyen-Congo, French Cameroon, Nigeria, the Gold Coast, Ivory Coast, Liberia and Sierra Leone." Storey (1930) had recorded a similar disease in East Africa. Working independently, both Storey and Golding showed that CMD could be transmitted by a white fly, *Bemisia* spp., now known to be *B. tabaci* (Russell 1957). Storey also showed that the disease was further spread in the field by vegetative propagation, from stakes of infected plants.

It was the widespread occurrence and high incidence of CMD that caused governments to start cassava-breeding programs aimed at obtaining cultivars resistant to the disease. Breeding programs were launched in Nigeria (Faulkner 1932), in Ghana (Miles 1934), in Tanzania in 1935 (Storey 1936), in the Congo, and in francophone Africa by Institut de recherches agronomiques tropicales et des cultures vivrières (IRAT). This paper will discuss the three plant-breeding programs that have used interspecific hybridization within the genus *Manihot* to obtain resistance to CMD.

Storey (1935) introduced "resistant" material from West Africa that proved susceptible under the conditions of his trials. He suggested that the West and East African strains of the virus were different, although it was not possible to show symptom differences between them. He demonstrated differences in the symptom expression of mosaic from separate sources in East Africa and postulated the existence of mild and severe strains of CMD. He was also able to demonstrate the existence of a second viral disease, cassava brown streak (CBS), which did not occur in West Africa (Storey 1936).

Until the early 1950s, hybridization and selection in the West African programs were confined to the species *Manihot esculenta*; these failed to produce any cultivars resistant to cassava mosaic. None of the cassava lines tested in East Africa showed high resistance to CMD or to CBS, and, in 1937, Storey first crossed cassava with ceara rubber (*M. glaziovii*), which appeared to be resistant to the disease. By 1939, Storey widened his interspecies crosses to include *M. dichotoma*, *M. cathartica*, and *M. dulcis*. He obtained a high level of field resistance to both CMD and CBS in the *glaziovii* × cassava and the *dichotoma* × cassava hybrids. Later still, *M. melanobasis* and *M. saxicola* were included in the program.

Storey used the term "field resistance," as do most plant breeders today. Field resistance is the sum of resistance of the plant to the virus per se and to infection from inoculation by the vector. Resistant plants are symptomless or show only occasional, mild leaf symptoms. Storey showed that none of his interspecies hybrids were immune to the virus, as he was able to infect field-resistant material by grafting.

This program was continued by Nichols and, later, by Jennings. Nichols (1947) reported that resistance to CMD was genetically controlled and that the *dichotoma* × *esculenta* interspecies hybrids showed the most promise for selection for mosaic resistance; these populations, however, showed a high degree of sterility. *M. glaziovii* × cassava interspecies hybrids and tree cassava × cassava hybrids both produced segregates resistant to CMD and CBS. In 1949 (EAAFRO 1949), it was shown that interspecies *M. melanobasis* × cassava hybrids had a high degree of resistance to CBS.

By 1951, the mosaic-tolerant ceara hybrids, notably 4332/2 and 46106/27, were available for testing and showed considerable promise (EAAFRO 1951). Jennings then started a program of intercrossing the resistant interspecies hybrids to release recessive genes for resistance and to recombine genes that had been dispersed in the earlier backcrossing program (Jennings 1976). This technique was successful, and considerable increases in stable resistance were obtained, notably in Jennings' clone 5318/34. This program was terminated in 1957; 91 clones were retained, 5 of which were selected and distributed to farmers. These improved clones, which are now grown on a limited scale in East Africa, have not replaced the established local cultivars (Bock and Guthrie 1978). The reason, according to Bock and Guthrie (1978), is that the new cultivars proved to be less palatable than the traditional types. In addition, farmers found that, as the health status of the new cultivars declined, their mosaic resistance appeared to break down and seemed in no way superior to that of other types.

When I began a breeding program in Nigeria, I selected eight clones on the basis of the severity of mosaic symptoms in the field; yield; and contrasting agronomic characteristics such as branching, plant height, and root arrangement. These varieties were crossed together in a diallel program that produced some 25 000 seedlings. None of this material was field resistant to mosaic (Beck 1959).

In my opinion (Beck 1971), adequate resistance to African cassava mosaic virus does not occur within the species *M. esculenta*. In 1956, seed populations from the East African program, which had been continued by Nichols and Jennings, were introduced to Nigeria. From seed derived from Jennings' selection 5318/34, I selected the CMD-resistant hybrid 58308 (Beck 1960); 58308 was selfed, intercrossed with other interspecies hybrids, and crossed to high-yielding local West African selections. In 1960, the cassava-breeding program was moved to the Federal Root Crops Research Station at Umudike in eastern Nigeria, and the progeny testing was continued by Ekandem. Unfortunately, almost all of the progeny and the records of this program were lost during the civil war in Nigeria. Most of the parent material, however, had been retained at Moor Plantation, and 58308 subsequently became the main source of resistance to CMD in the breeding program at the International Institute of Tropical Agriculture (IITA). The hybrid 58308 has remained resistant to CMD for 22 years under high-inoculation pressure (Hahn et al. 1980).

IITA commenced a cassava-breeding program in 1971. About that time, a new and serious disease of cassava was reported in West Africa. This disease, cassava bacterial blight (CBB), is caused by the bacterium *Xanthomonas manihotis*. The disease was first recorded in Brazil (Bondar 1912) and was first reported in Nigeria in 1972 (IITA 1972). Since that time, serious outbreaks of CBB have been reported in Zaire, Cameroon, Togo, Benin, Ghana, Congo, Uganda, Kenya, Tanzania, Burundi, Rwanda, and Central Africa (Terry 1978c). Most indigenous African cultivars have proved susceptible to the disease; infection can result in complete crop loss. Thus, the IITA program focused on both CMD and CBB. It was found that the interspecies hybrid 58308, originally selected for resistance to CMD, was also highly resistant to CBB.

Jennings, using the method of Gilbert (1967), showed that, in the East African material, resistance to both CMD and CBB was inherited in a predominantly additive manner, although there was a significant specific combining ability effect for resistance to CMD. Hahn and Howland (1972) confirmed this finding as well as showing that CMD resistance was under polygenic control and was recessive, with a heritability of about 60%.

This relatively high heritability, along with the polygenic control, meant that the resistance was likely to remain stable. Furthermore, a comparison of nine isolates of the CBB pathogen on resistant clones at IITA (1977) suggested that the resistance to CBB was race-nonspecific (Parlevliet and Zadocks 1977).

Hahn et al. (1980) showed that progenies derived from 58308 crossed with susceptible and moderately susceptible clones exhibited a significant genotypic correlation between resistance to CMD and resistance to CBB ($r = 0.90$). The suggestion was that the correlation reflected links between the genes for resistance, controlled by a number of linked loci on a chromosome or on a set of chromosomes.

These workers have used the technique of population improvement with a half-sib family selection scheme, using relatively large families to break the linkage of the resistant genes and those controlling the poor agronomic characters of 58308, and to incorporate the resistance genes or groups of genes with those for high yield, good quality, low-cyanide content, and resistance to lodging. Because of the efficient field selection carried out by Hahn and his co-workers, this method has been brilliantly successful. Rapid improvements have already been made, and improved selections are in trial over a wide range of environments.

Resistance to the cassava green mite (CGM) *Mononychellus tanajoa*, a new pest in East Africa, has also been identified in these populations in trials in Tanzania.

Because of the rapid distribution of the IITA progeny for field testing by national programs, further information will be obtained on the stability of resistance to CMD and the race specificity of resistance to CBB. Efficient local selection within the IITA populations will lead to the development of well-adapted types either to be distributed to farmers or to act as parent material in national cassava-improvement programs.

Discussion

All the work described in this paper has been aimed primarily at producing cassava cultivars adapted to local African farming conditions for use as human food. In most African farming systems, cassava is grown by smallholders in fields less than 5 ha. It is harvested as required by the farming family, either for their own use or for processing as a cash crop into products such as chips, flour, or gari. There are marked producer and consumer preferences as to the type of cultivar grown, and these probably determine the uptake of new cultivars in peasant societies. For example, in Kenya, the mosaic-susceptible cultivar F279, introduced by Storey in his breeding program, has become popular in local areas, but the mosaic-resistant, improved cultivars are nowhere grown on a large scale and have not replaced the local types. Consumer preference is a local phenomenon and may be expected to limit the widespread use of otherwise outstanding cultivars.

However, the uses of cassava are now expanding. The recent increases in oil prices have made the development of synthetic fuels economically attractive. Zimbabwe and Kenya have commenced ethanol production programs using the molasses by-products of their sugar industries as a fermentable carbohydrate source. These, and other, governments are examining the possibility of using hydrolyzed cassava starch as a feedstock for synthetic fuels. Cassava is capable of giving a higher yield of calories per hectare than is any other crop. The mathematics of the utilization of cassava are attractive: 1 t fresh cassava roots will produce 180 L ethanol fuel compared with 70 L from 1 t sugarcane.

Kenya and other African states are also developing livestock-feed industries based on cassava chips or cassava pellets, both for internal use and for export to Europe. The economics of using a food crop as an industrial feedstock, however, require careful examination.

To produce enough cassava to supply factories will require large-scale, plantation-type agriculture, and this type of cultivation requires mechanical assistance. Also, the types of cassava developed for large-scale agriculture will be markedly different from those adapted to peasant agriculture. For example, these cultivars will require much higher levels of disease resistance, because large-scale monocropping systems are more vulnerable to epiphytotic outbreaks of disease and, thus, more economically serious crop losses.

In my breeding programs, I sought to develop types that quickly formed a closed canopy and consequently reduced weeding. A minor heresy was invoked, for, in these closed canopies, susceptibility to *Cercospora henningsii* was an advantage. In Nigeria, this pathogen attacked only the older leaves in the lower strata of the canopy during the wet season and, in effect, hastened leaf fall; because they were densely shaded, these leaves contributed little to carbohydrate production and probably were respiratory sinks for sugars. *Cercospora* acted as a useful pruning agent. These cultivars would be of limited use in plantation agriculture. Densely interwoven branches in the canopy may cause serious problems in large-scale harvesting operations, and to use cultivars with no resistance to a potentially dangerous pathogen would be an unacceptable risk. A more erect, sparsely branched type, easy to clear from the fields, would be required. These types have a more limited leaf area; thus, resistance to fungal leaf diseases would be of considerable importance. Short, thick roots growing close to the original planting material would be essential for mechanical lifting. To enable harvests to be staggered for a continuous supply to the processing plant, it is essential that industrial cultivars have roots that are resistant to rotting when mature.

Most cassava-breeding programs have been designed to select cultivars adapted to the low-input systems of agriculture used by smallholder farmers. It is probable that large-scale, continuous cultivation for industrial production will require the use of fertilizers to maximize yields. The new cultivars must have the genetic background to permit economic responses to improved nutrition. In producing these cultivars, the breeders' work will be simplified in at least one regard: they will be selecting for an objective goal, fermentable carbohydrate per hectare, and will not be bedeviled by subjective considerations of local consumers' tastes.

The successful breeding programs of Hahn and his colleagues at IITA have provided a range of excellent commercial cultivars and sources of resistance to the major diseases and pests of the crop. The distribution, availability, and use of this valuable material are seriously restricted by the necessary plant-quarantine regulations of all cassava-growing countries. Successful genotypes have been a long time in the building. It has now been shown (IITA 1978) that the heat treatment for cuttings developed by Chant (1959) does not guarantee disease-free material, and, thus, the distribution of standard vegetative material is not possible. Persley (1979) has shown that CBB can be transmitted through the seed. The use of thermal inactivation of the bacterium by hot-water treatment of the seed requires further investigation before the plant-quarantine authorities will allow unrestricted movement of heat-treated seed. At present, most plant-quarantine stations do not have sufficient facilities to permit the growth and quarantine of the large seedling populations required by plant breeders. Perhaps the most promising technique for the rapid dissemination of improved cultivars is by the tissue-culture methods of Kartha and Gamborg (1975), coupled with the single leaf-bud cutting technique of Roca et al. (1980). The observation of Bock and Guthrie (1976) that CMD may be latent in vegetative material and may not produce symptoms until the third cycle of propagation should be borne in mind. This type of symptom masking will most likely be in cultivars with a high level of resistance to CMD.

Increasing population pressure on land and rapidly declining soil fertility make the rapid dissemination of IITA's improved cassava genotypes of paramount importance. The need is most urgent in areas where whitefly populations and the levels of infection with CMD and CBB are consistently high. A good case can be made for the establishment of an intermediate quarantine station, possibly outside Africa, using the new techniques of chemical and thermal inactivation of seed-borne pathogens and tissue-culture and rapid-propagation techniques on a large scale. It is on the ready availability of improved genotypes that national programs will depend in future. The need for them is now.

Research priorities, techniques, and accomplishments in cassava breeding at IITA

S.K. Hahn[1]

The major biological constraints to cassava production in Africa are diseases, especially cassava mosaic disease (CMD) and cassava bacterial blight (CBB), and insect pests, such as the mealybug and the green mite. Thus, the primary objectives of cassava breeding at IITA have been to alleviate these constraints. Cassava-breeding techniques such as pollination, seed collection, seed germination, and screening for important agronomic characteristics are used; the procedures and yield trials are described in this paper. The most promising IITA cultivars with good performance in a wide range of environments within Nigeria are TMS 30572 and TMS 30555. The most CMD- and CBB-resistant cultivar is TMS 30001, which is also low in cyanide. Sources of resistance to cassava green mite have been identified, and a mealybug-resistant *Manihot* species has been found. The improved populations and families of true seeds with sources of disease and insect resistance, high-yield potentials, compact roots, and low-cyanide levels are available for selection.

En Afrique, les plus importants obstacles biologiques à la production du manioc sont les maladies, surtout la mosaïque et la brûlure bactérienne, et les insectes comme la cochenille et la teigne. La sélection génétique du manioc à l'IITA a donc eu pour objectif principal de lutter contre ces ennemis. Les techniques de sélection, telles la pollinisation, la collection et la germination des semences, et l'introduction des propriétés agronomiques importantes sont utilisées ; l'article décrit les méthodes et les essais de production. TMS 30572 et TMS 30555 sont les cultivars de l'IITA les plus prometteurs en termes de bon rendement dans diverses régions du Nigéria. TMS 30001 à faible teneur en cyanure, résiste le mieux à la mosaïque et à la brûlure bactérienne. On a identifié les facteurs de résistance à la teigne et on a trouvé une espèce de *Manihot* résistante à la cochenille. On dispose maintenant, aux fins de sélection de population et de familles améliorées, de semences résistantes aux maladies et aux insectes, ayant un rendement élevé, des racines fermes et peu de cyanure.

The annual African production of 42×10^6 t of cassava, grown on 6×10^6 ha, constitutes 40% of the food production from 55% of the total area under cultivation (FAO 1977). Cassava provides more than 50% of the caloric requirements for 200 million Africans. And yet yields are low in Africa: about 7 t/ha in 12 months, compared with potential yields of more than 20 t/ha.

The major biological constraints to cassava production in Africa are diseases, especially cassava mosaic disease (CMD) and cassava bacterial blight (CBB), and insect pests such as the cassava mealybug and the green mite (Hahn et al. 1979). CMD causes yield reductions of 20–90%, and CBB causes complete loss of tuberous root yield and leaves during severe epiphytotics in many countries in Africa. Cassava mealybug (CMB) has now been reported from most of the major cassava-growing countries in Central and West Africa, its presence being first reported in Zaire in 1973 (Hahn and Williams 1973). The cassava green mite (CGM) has also become a serious pest of cassava in Africa since its presence was first recorded in Uganda in 1972 (Nyiira 1972). It is believed that both CMB and CGM were introduced from Latin America.

Cassava-breeding procedures

IITA's cassava-breeding objectives, which were established on the basis of production constraints, are to develop clones that are

[1] International Institute of Tropical Agriculture (IITA), P.M.B. 5320, Ibadan, Nigeria.

high yielding, are resistant to major diseases and insect pests, are adaptable to a wide range of environmental conditions, and possess high-quality characteristics. To achieve these objectives, the current breeding scheme (Fig. 1) was adopted.

Source populations

The first step was to acquire a source population with high frequencies of genes associated with desirable characters. The base material is local and introduced germ plasm and improved populations and families. Hybridization is used as a means to acquire and improve the source populations. Parents for crosses are selected for their performance and breeding values.

Pollination

Hand pollination Pollen is collected early in the morning (before 10:00), and pollination is done later the same day. Both male and female flowers that are about to open are used. After removal of male flowers, female flowers are bagged in fine cloth or white paper bags before pollination. When the anthers are mature, they change from green to yellow.

Pollination is often done by hand with a stick, the tip of which is covered with a piece of velvet-like material to which the pollen readily adheres. Several flowers can be pollinated before the applicator needs to be recharged. If the same applicator is to be used for other pollen parents, it is dipped in alcohol before reuse. The pollinated flowers are then placed in cloth or paper bags (white) to stop bees or other insects from introducing foreign pollen.

Pollination in isolation Because cassava is normally a cross-pollinated plant, outcrossing can be done among the selected parents in isolation. The isolation plot is set up with several selected parents, with single or a few plants (up to three) per parent, and with several replications to provide an equal chance of crossing.

Seed collection

The seed matures about 70–90 days after pollination. Fruits, collected when the coat begins to shrivel, are dried in the sun or in an oven at 40–50°C until the fruits shatter. Fruits from an isolation plot are collected in cloth bags that are hung on cassava plants and left there until the fruits inside shatter.

Cassava seed has either a short or no dormant period. Comparative tests of 0–7-year-

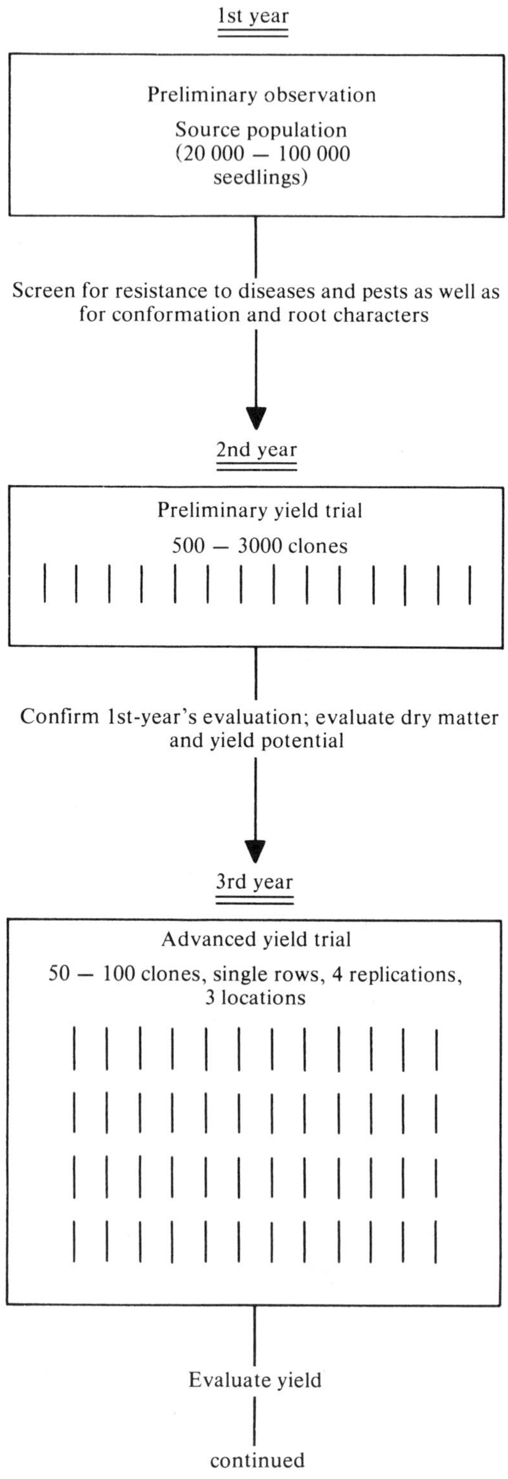

Fig. 1. concluded

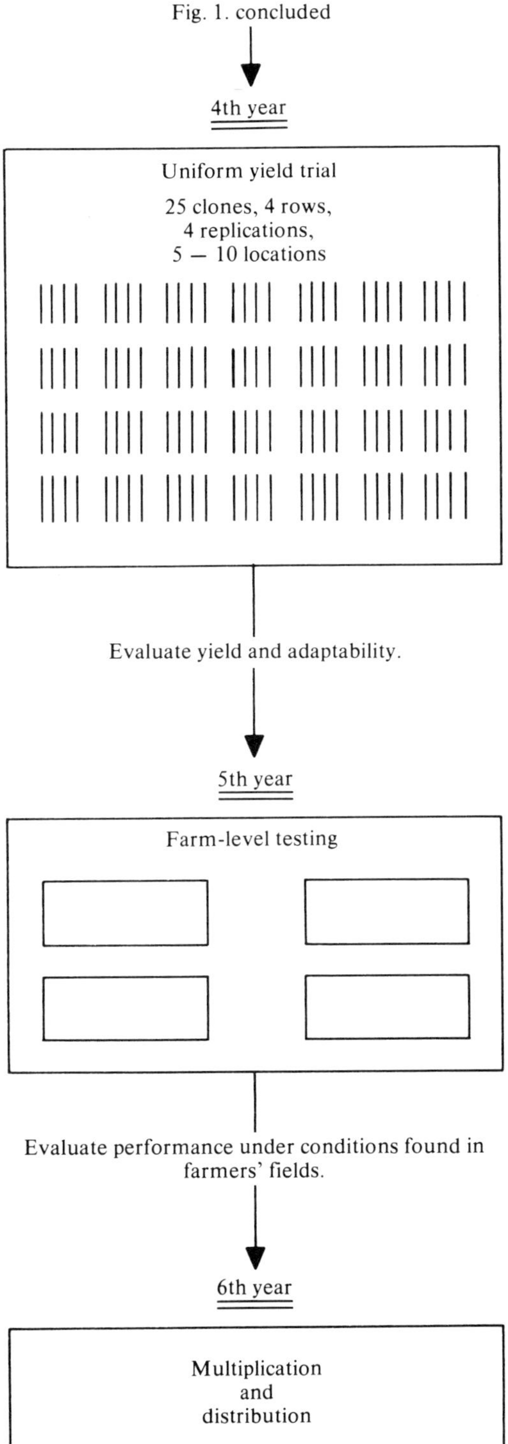

Fig. 1. *IITA cassava-breeding procedures.*

old seed, stored at 5°C and 60% relative humidity, showed no significant differences in viability.

Seed germination

The seed germinates quickly at optimal soil temperature (30–35°C) and moisture treatments (irregular irrigation). Scarification is not necessary. At IITA, seed is planted in the dry season (January and February), and seedbeds are irrigated daily for about 3 weeks and at regular intervals until the rainy season commences. The seeds take about 2–3 weeks to germinate. During this period, weeds are a problem. Trifluralin (1.4 kg/ha) mixed with paraquat at 0.5 kg/ha provides good weed control. Spray application of paraquat 10 days after seeding controls weeds until the seedlings are big enough for hand weeding.

Seeds can be pregerminated in the greenhouse for use in areas where soil temperature is below that required for germination and where irrigation facilities are not readily available. Each seed is planted in a peat pellet, jiffy pot, or plastic bag, and the seedlings are planted in the field when they are 30 cm high.

Selection

The seedlings are screened for resistance to diseases and insects at 3 and 6 months after planting. They are exposed to a large population of whiteflies (vector of CMD) from spreader varieties. If environmental conditions are favourable for disease development, seedlings are also screened for resistance to CBB under natural epiphytic conditions. If not, seedlings are inoculated with bacteria through the stem-puncture method. Screening for CMD and CBB is based on the system proposed by IITA. Susceptible seedlings are eliminated. At this stage, seedlings from low-cyanide populations are selected for cyanide content by the picrate method. The selected seedlings are uprooted after 1 year and screened for conformation and root characteristics. Low-branching types (branching heights of about 50–100 cm) are discarded. The seedlings with short necks (1–3 cm) and uniform, short, compact, fat roots are selected.

In the 2nd year, the selected seedlings (about 500–3000) are cloned and planted for preliminary yield trials in a single-row plot containing 3–5 plants per plot, with single replication and planting at 1 m^2 spacing. A standard variety is also planted every 10 clones for comparison. At this stage, the evaluation from the 1st year for disease and conformation is confirmed for each clone. The clones that perform poorly in terms of establishment, growth, and resistance to diseases and insects are discarded. Only the selected clones are evaluated for dry matter, yield potential, and other quality characters. Clones selected to be low in cyanide are evaluated quantitatively for cyanide content by an enzymatic assay. (The leaf cyanide score — tested by the picrate method — has been shown to be closely related to root cyanide content as determined by the enzymatic assay method.)

In the 3rd year, the best 50–100 clones selected in the preliminary yield trials in the 2nd year are used for advanced yield trials with single-row plots (10 m long) and four replications. These trials are conducted in three locations that cover a wide range of environments. Planting space is 1 m^2. At this stage, the clones are evaluated for yield, resistance to diseases and insects, root characteristics, conformation, dry matter, and consumer acceptance.

In the 4th year, the most promising 20 clones from the advanced yield trials of the previous year are advanced to uniform yield trials in 5–10 locations with four-row plots (10 m long) and four replications. These clones are evaluated for yield, consumer acceptance, and adaptation.

In the 5th year, the elite five clones from the uniform yield trials are planted for farm-level testing and farmer evaluation.

In the 6th year, the clones most popular with the farmers are multiplied and distributed for popularization of the varieties.

Accomplishments

The most promising cultivars, to date, are TMS 30572 and TMS 30555. From 1973 to 1978, these two showed superior performance, with average yields of about 40 t/ha in 1 year without fertilizer and adaptation to 26 environments in Nigeria, including locations with different soil, climatic, etc. conditions and rainy and dry seasons. They are moderately resistant to mosaic disease and bacterial blight. These cultivars have about 30% dry matter and good consumer acceptance. They require comparatively little weeding because of their good canopy, and they have relatively good root shape. Their progeny have performed well in Nigeria, Sierra Leone, Liberia, Gabon, Zaire, Zanzibar, Seychelles, and India.

TMS 30001 and TMS 30395 are very resistant to both CMD and CBB (Terry 1978a), give reasonable yields, and have acceptable consumer quality. TMS 30001 has high branching characteristics and may be suitable for intercropping with companion crops.

TMS 30001 is also low in cyanide (< 5 mg/100 g fresh weight), and others have been identified. Some, like TMS 30001, are resistant to both CMD and CBB. Consumer acceptance is good, and yield potentials are reasonable. These low-cyanide cultivars were selected from more than 300 000 genotypes tested in the last 6 years. Using the semiquantitative picrate method at the initial screening stage, researchers selected the most promising low-cyanide cultivars and quantitatively tested them using enzymatic assay, which has been reported to be most accurate.

CGM-resistant cultivars were selected from IITA material by Allen Carpenter and his colleagues in Zanzibar. They estimated that, if the yield could be sustained on a country-wide scale at 30% of the yield they had obtained, an increased production of a quarter million tonnes over pre-mite levels could be anticipated. The cultivars were acceptable in consumer quality. IITA has also identified cultivars that are resistant to the pest. The source of resistance is available from groups in Zanzibar, Tanzania, and IITA.

A *Manihot* species introduced in seed form from Brazil has shown resistance to the cassava mealybug (*Phenacoccus manihoti*). Hybrid seeds of the species with the cultivated cassava (*M. esculenta*) were obtained. Limited quantities of seeds from the resistant species or of the hybrid will be available soon.

Six improved populations of true seed with sources of disease and insect resistance, high-yield potentials, good root shape, and low-cyanide levels are available for selection. The populations have wide genetic variation incorporating more than 2000 Latin American and Asian sources. A large number of families that have proved to be good in breeding for important agronomic traits are also available.

Research priorities, techniques, and accomplishments in sweet-potato breeding at IITA

S.K. Hahn[1]

Sweet potato is an important staple-food crop in many parts of the tropics and subtropics. It is grown over a wide range of environmental conditions and has a tremendous capacity to produce dry matter per unit of land and time. The major biological constraints to sweet-potato production in the tropics are weevils and viruses. The primary objective of sweet-potato breeding at IITA, therefore, has been to improve sweet-potato clones and populations resistant to weevils and viruses. The sweet-potato-breeding methods and procedures are described in this paper, with particular emphasis on performance evaluation of breeding material. The IITA-improved sweet-potato cultivars with the most weevil-resistant roots and virus resistance are reported. Cultivars with improved agronomic characters, such as keeping quality, dry-matter content, and drought tolerance are also reported.

La patate douce est un aliment de base important dans plusieurs régions tropicales et subtropicales. On peut la cultiver dans divers environnements et elle fournit une quantité impressionnante de matière sèche par superficie et période de culture. Les principaux obstacles biologiques à la production de la patate douce dans les tropiques sont les charançons et les viroses. La sélection génétique de la patate douce à l'IITA a donc eu pour objectif premier d'améliorer les clones et les populations de patate douce résistants à ces deux ennemis. L'exposé décrit les méthodes et les procédés de sélection de la patate douce et met l'accent sur l'évaluation du rendement du matériel génétique. On mentionne les cultivars de patate douce améliorés de l'IITA ayant les racines les plus résistantes aux charançons et la meilleure résistance aux viroses, ainsi que les cultivars possédant des propriétés agronomiques améliorées comme l'aptitude à la conservation, le contenu de matière sèche et la résistance à la sécheresse.

Sweet potato (*Ipomoea batatas*) is the sixth most important food crop in the world, with an annual production of about 1.4×10^7 t from 1.5×10^7 ha (FAO 1977). Although it is produced more in the temperate zone than in the tropics, it is grown widely and is an important staple-food crop in many parts of the tropics and subtropics. Both tuberous roots and leaves are eaten, the former mainly as a source of carbohydrate and the latter as a source of protein and minerals. Both parts have substantial amounts of vitamins, particularly vitamin A. In addition to its importance as a human food, sweet potato provides animal feed and raw material for industrial purposes.

It has a tremendous capacity to produce dry matter per unit of land and time. The caloric production from the edible root of sweet potato in Japan is 2.94×10^7 kcal/ha (an average root yield of 2.1×10^4 kg/ha in 5 months). This is almost five times that of rice, which has an average yield of 4.5×10^3 kg/ha (Murata et al. 1976). It is grown over a wide range of environmental conditions, between latitudes 40°N and 40°S and between sea level and 2300 m. Even on land of low fertility and relatively low pH, it can still produce a considerable yield (Hahn 1977a; Hahn and Hozyo 1980). It requires low production inputs compared with other food crops and is less vulnerable to adverse conditions such as drought and heavy storms. For these reasons, it provides food at a time when other staple foods are in short supply.

[1] International Institute of Tropical Agriculture (IITA), P.M.B. 5320, Ibadan, Nigeria.

The major biological constraints to sweet-potato production in the tropics are weevils and viruses. Yield reduction from viruses is reported to be 78–80% in Nigeria (Hahn 1975, 1979) and 57% in Uganda (Mukiibi 1977). Weevils cause damage to both tuberous roots and leaves in the field, as well as reducing quality, storability, and market value. Storage is also a problem in the tropics because of rapid physiological losses brought about by high temperatures, serious weevil damage, and storage rots. Likewise, preservation of planting materials is a problem in areas of the tropics with a long dry period.

These constraints have seriously limited production; yet they received little attention before 1971 when the International Institute of Tropical Agriculture (IITA) initiated sweet-potato-breeding work. The Institute's objectives were, and still are, to produce cultivars that are capable of high yield in terms of dry matter per unit of land and time; that are resistant to economically important diseases and insects; and that have good keeping qualities, improved consumer acceptability, processing characteristics, nutritional value, and wide adaptation. For the past 8 years, particular emphasis has been placed on breeding improved cultivars that are resistant to weevils and viruses.

Breeding procedures and activities

The sweet-potato-breeding scheme that has been adopted was designed for tropical conditions, especially the prevailing climatic and biological conditions at IITA (Fig. 1).

Sweet potato is normally a cross-pollinated (by insects), vegetatively propagated crop. The recurrent selection system used for population improvement is half-sib family testing and selection between and within families. Mass selection for resistance to viruses and weevils is also being used. This system is aimed at increasing population means and retaining a high degree of genetic variability through continuous, cyclic evaluation, selection, and recombination.

The short-range program for sweet-potato improvement focuses on developing cultivars with resistance to viruses and weevils; the long-range program is more generalized and demands more genetic variability. This two-pronged improvement approach is appropriate for IITA because it allows the Institute to

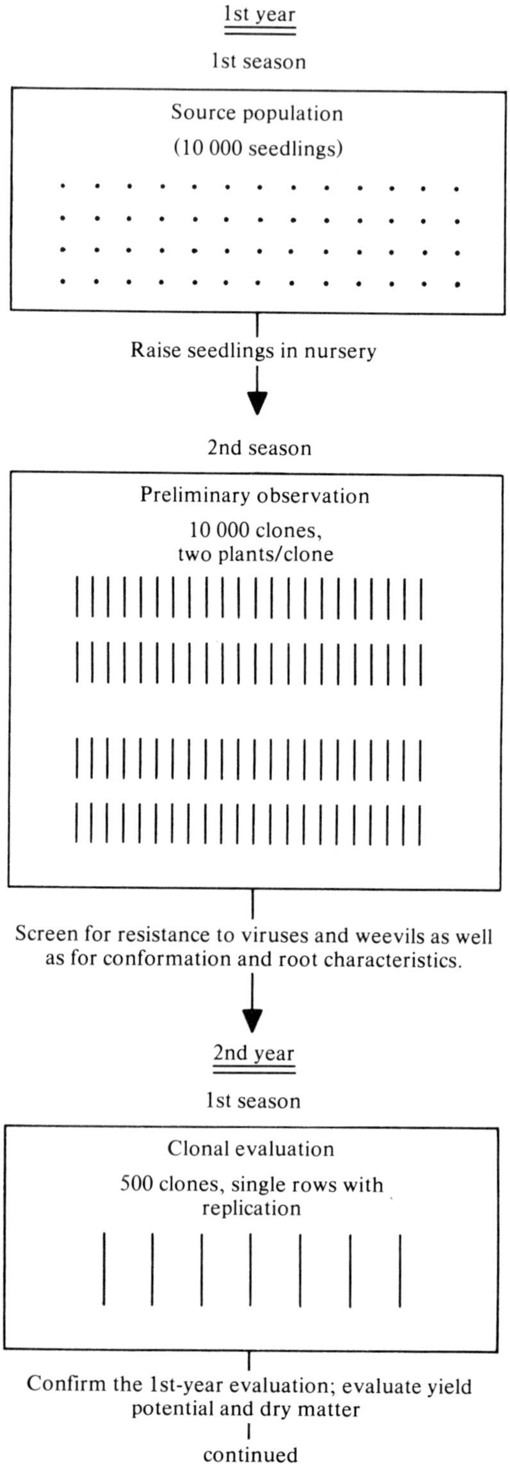

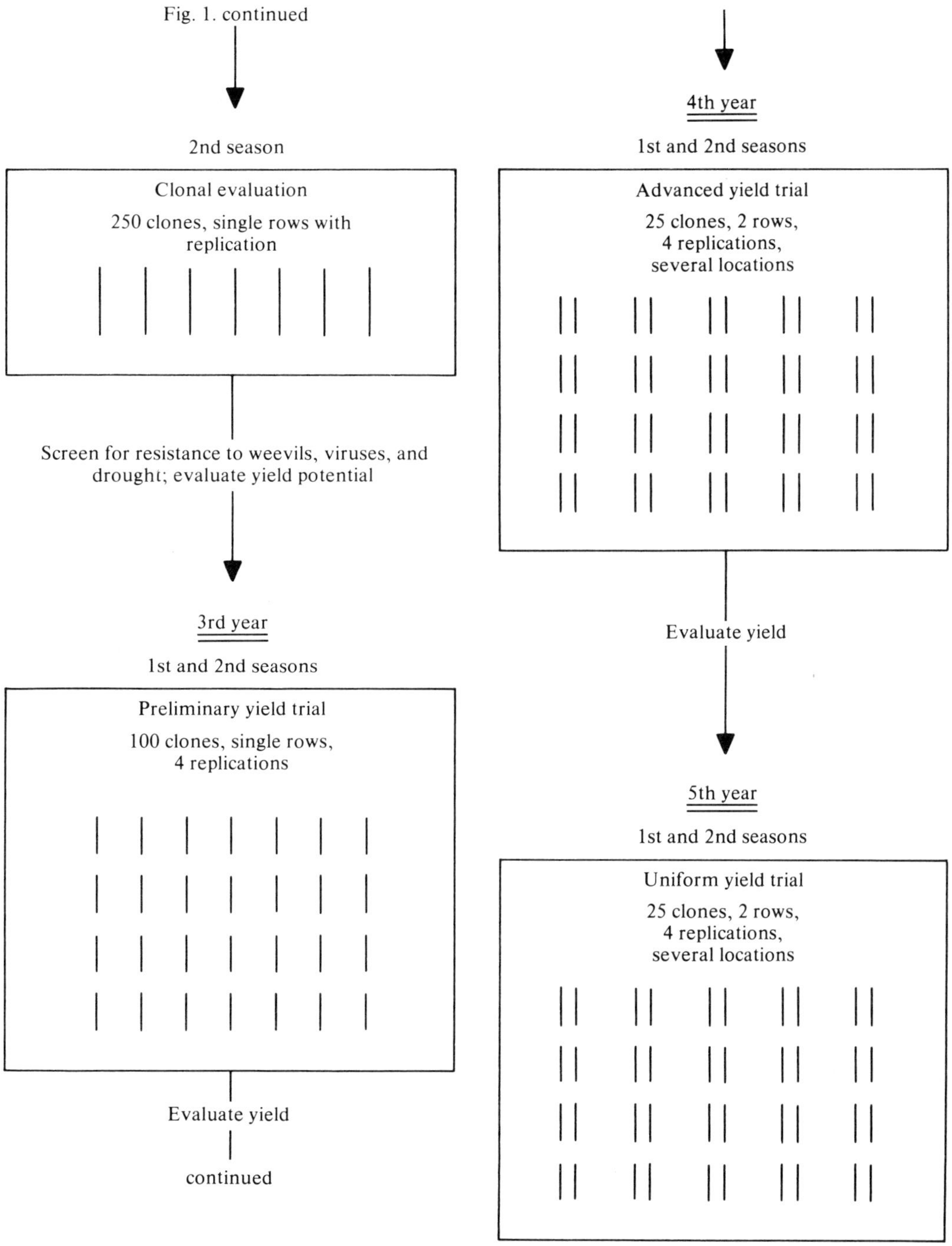

Fig. 1. *Sweet-potato-breeding procedures at IITA.*

serve many national programs, providing them with genetically diverse seed for selection under local conditions. In addition, the conventional intervarietal crossing program is also used to meet specific needs.

Source populations

When IITA began its improvement program in the early 1970s, the first step was to acquire the source populations from which genotypes with desirable characters could be selected. The base material now includes: local and introduced germ plasm and improved populations and families. It was initially difficult for IITA's sweet-potato-breeding program to introduce germ plasm in a vegetative form from outside Nigeria. As an alternative, germ plasm in seed form has been introduced from many countries in Africa, the Americas, and Asia. Nigerian local cultivars were also used and were the starting point for the breeding program. Several clones, selected from the source populations, showed good performance and high breeding values for important traits and were intercrossed in isolation to make new source populations.

Selection

In the 1st season (rainy season) of the 1st year, about 10 000 seedlings are raised in the nursery and are moved to the field 2 months before the end of the rainy season. During the next few months, they are screened for resistance to viruses and weevils and for conformation and tuber characters.

In the 1st season of the next year, the selected clones (about 500) are cloned and planted for observation in a single-row plot 4 m long, with 1 m between rows and 30 cm between plants within rows. A standard variety is also planted every 10 clones for comparison. At this stage, the 1st-year evaluation for resistance to viruses and weevils and for conformation and root characters is confirmed. About 250 clones that are good in terms of establishment, growth, and resistance to viruses and weevils are selected.

In the 2nd season of the year, the selected 250 clones are replanted and screened for resistance to weevils and viruses, drought tolerance, and for yield potential. The most promising 100 clones are selected.

In both seasons of the 3rd year, a preliminary yield trial is conducted for the 100 promising clones. They are planted in single-row plots with four replications. Between-plant spacing is 30 cm within the 10-m long rows that are 1 m apart. At this stage, a more thorough yield evaluation is done in addition to testing for resistance to viruses and weevils and for dry-matter percentage and keeping quality.

In both seasons of the 4th year, the best 25 clones selected from the preliminary yield trial are put forward for advanced yield trials in two-row plots (10 m long, 1 m apart) with four replications. Advanced yield trials are conducted in three locations, covering a wide range of environments. At this stage, more formal performance tests are done for yield, resistance to weevils and viruses, dry-matter percentage, consumer acceptance, adaptation, and keeping quality.

The next year, the most promising 25 clones from the advanced yield trials are advanced to uniform trials in four locations within Nigeria with two-row plots (10 m long, 1 m apart) and four replications. These clones are evaluated for yield, consumer acceptance, keeping quality, and adaptation.

In the 6th year, the elite five clones from the uniform trials are planted for farm-level testing and farmer evaluation. From the 7th year, the clones that are most popular with farmers are multiplied and distributed for popularization.

Accomplishments

The IITA-improved, sweet-potato cultivars with the most weevil-resistant roots are TIS 3053 and TIS 3030 followed by TIS 2154, TIS 2153, TIS 3290, TIS 3017, TIS 2532, and TIS 2544. The cultivars with the most resistant shoots are TIS 2532, TIS 3017, and TIS 3030 followed by TIS 1499 and TIS 1487 (IITA 1978). These clones are high yielding (13–30 t/ha) in Nigeria, well accepted by consumers, white fleshed, dry in texture when cooked, and have good keeping qualities. They also exhibit resistance to viruses; however, the IITA-improved cultivar that shows the most resistance to viruses is TIS 2498 followed by TIS 3053, TIS 3030, TIS 2532, and TIS 3017 (IITA 1979; Terry 1979). These improvements in resistance have been complemented by work on underground-storage systems, which have proved effective against weevil attack (IITA 1979).

Sweet-potato improvement in Rwanda

M.J.J. Janssens[1]

Sweet potato is a staple food throughout most of Rwanda, but production is hampered by viruses, weevils, leaf caterpillars, environmental heterogeneity, the late maturity of the crop, intergenotypic competition, and taste requirements. These constraints are the focus of the sweet-potato-improvement program in Rwanda. A main breeding population was established for evaluation of yield potential under mid-elevation Rwandan conditions, and subpopulations were used for trials on drought and cold stress. The procedure comprised selection, preliminary yield trials, multilocality yield trials, and preextension trials. In trials so far, tuber yield was highly correlated with number of tubers, and top growth was an indicator of yield quality (high starch and dry-matter content). Saccharose and starch were negatively correlated. One superior clone, Rusenya, yielded 8.1×10^3 kg/ha on average across seven environments. Groups of varieties with similar environmental adaptability could be identified. A system of multireciprocal recurrent population improvement is proposed that consists of a central breeding population, aimed at high and stable yield, and a set of interdependent breeding subpopulations corresponding to particular breeding goals.

La patate douce est un aliment de base dans la majeure partie du Rwanda, mais la production est limitée par des virus, charançons et chenilles, l'hétérogénéité environnementale, la maturité tardive des récoltes, la concurrence intergénotypique et des problèmes de goût. Le programme d'amélioration de la patate douce au Rwanda vise à corriger cette situation. On a constitué une population principale pour évaluer la productivité potentielle en moyenne altitude, et des sous-populations pour vérifier la résistance à la sécheresse et au froid. Le processus comprenait la sélection, des essais de rendement dans plusieurs localités, des essais préliminaires de rendement, et de vulgarisation. Jusqu'à maintenant, les essais ont montré une forte corrélation entre le rendement des tubercules et leur nombre, et la croissance maximale a été un indicateur de la qualité du rendement (riche en féculents et en matières sèches). La saccharose et les féculents étaient en corrélation négative. Un clone supérieur, le Rusenya, a donné un rendement moyen de $8{,}1 \times 10^3$ kg/ha dans sept environnements. On a pu indentifier des groupes de variétés ayant les mêmes facultés d'adaptation à l'environnement. On propose un système d'amélioration de population récurrente multi-réciproque constitué d'une population centrale devant atteindre un rendement élevé et stable et de sous-populations interdépendantes répondant à des objectifs précis de sélection.

Sweet potato is the most important root crop in Rwanda. It is grown on about 1.5×10^5 ha and produces an average 7–8 t/ha. In the past, sweet-potato research for Rwanda, Burundi, and Zaire was mainly conducted at a research station in Zaire (Mulungu). Breeding focused on virus resistance, cold tolerance, and protein quality and led to the release of such clones as M46, D.Virovsky 16, and variety 5037 (Lemarchand 1956).

In Rwanda, research on sweet potato was restricted to introductions, preliminary yield trials (Camerman and van Bellinghen 1973), and comparative yield trials, with subsequent release of some of the clones. Only in 1978 did an active research program of sweet-potato improvement and breeding begin at Rubona (Institut des sciences agronomiques du Rwanda, ISAR).

Viral diseases constitute a major production problem (Lemarchand 1956). During the last 2 decades, ISAR released several clones (Anne-Marie, Mugenda, and Caroline Lee), all of which turned out to be virus susceptible. The pioneer work of Sheffield (1957) identified two strains of virus affecting sweet potatoes in Uganda. However, the wide range of

[1] Institut des sciences agronomiques du Rwanda (ISAR), B.P. 138, Butare, Rwanda.

virus-like symptoms in Rwanda fields suggests that a greater number of strains affect production here.

A second major production constraint is weevils, *Cylas* spp. (Buyckx 1962). Weevils are particularly a problem for late plantings (April–May) on hill sites, because their populations expand rapidly during the dry season (July–September).

The leaf caterpillar, *Acraea acerata*, also can become a serious production problem during the dry season (Lefevre 1948). Erinose, characterized by abundant pubescence, is a minor pest caused by *Aceria* spp. (Sheffield 1962).

Alternaria anthracnosis is a minor disease problem (Buyckx 1962). Clones Nyiramujuna and Rukocoka, as well as their offspring, appear very susceptible, particularly under cold and humid conditions.

Although Rwanda is only about 2.8×10^4 km^2, no fewer than 12 agroecological regions can be distinguished (Delepierre 1974).

Cultural practices, too, are diverse and need to be better understood in both agronomic and economic terms.

Most of the local varieties of sweet potato mature between 6 and 8 months. However, the increasing demographic pressure in Rwanda is forcing producers to harvest the crop before maturity. Release of some early-maturing clones (about 4–5 months) would improve yields.

Consumer taste also deserves attention because, ultimately, it determines whether a new variety will become an important element in people's diets.

Improvement program

When the sweet-potato-breeding program was started in 1978, about 40 clones were available in Rubona — half of them local introductions and half, accessions from Zaire. Among the latter, many clones could be traced to obsolete US varieties, e.g., Caroline Lee and Red Jersey.

It was decided to generate genetic variability as quickly as possible by the introduction of seed material from abroad and clonal material from the different regions of Rwanda, as well as intercrossings of all 40 clones of the existing collection.

A progressive, nested polycross design (Janssens 1980) was used. Four sets of seven seed gardens were established. In the first set, a single recurrent parent was used for pollination of each of the seven clones; in the second set, two recurrent parents were used for pollination of each of seven other clones; in the third set, the mix contained three recurrent parents; and, in the fourth set, four pollen parents were used. A total of 10 pollen parents was used, chosen either for high-yielding capacity or for other desirable characters, e.g., early maturity or orange flesh. Hence, 38 (10 + 28) clones were intercrossed in 28 seed gardens.

The isolation distance between seed gardens within each set was about 100 m, which is far below the recommended 500 m for total isolation (Hahn, elsewhere in this volume). However, part of the contaminating pollen mix within each set is desirable. For example, a contamination of 20% in the biclonal seed gardens (i.e., first set of seed gardens with only one single recurrent pollen parent) will actually correspond to an undesirable contamination of only 10%, as the other half is desirable pollen of the recurrent pollen parent. In the fourth set, a 20% total contamination reduces to a mere 4% undesirable contamination.

Unfortunately, the flowering habit of the 38 clones was very heterogeneous. In future, this could be overcome by the use of some flowering-inducing methods, e.g., grafts of flowering clones (AVRDC 1976); some suitable root-stock species (yet to be found, for Japanese morning glory, *Pharbitis nil*, is unadapted to the Rwandan short photoperiod); application of appropriate pesticides (Jones 1977); use of phytohormones (Howell and Wittwer 1955; Suge 1977); nitrogen fertilization (Knavel and Lasheen 1969); or any suitable combination of these methods. If flowering could be controlled in a reasonable way, the progressive, nested polycross design would provide estimates of both the additive and dominance variance components through regression analysis (Janssens 1980).

One main breeding population was constituted, comprising high yielders known to be well adapted to the central mid-elevation region of Rwanda (1500–1900 m). In addition, two breeding subpopulations were constituted, one adapted for the high elevations (1900–2300 m) of northern Rwanda and the other adapted to the drier conditions (600–1000 mm annual rainfall) and lower elevations (1250–1500 m) of eastern Rwanda.

Selection procedure (Table 1)

Genetic variability is initiated through introductions and crossings. After scarification and pregermination, seedlings are raised in the nursery at a 10 cm × 5 cm spacing for about 2–3 months, up to the 4–6 leaf stage. For the high- and mid-elevation programs, the seedlings are transplanted to the selection fields (SFs); however, for the east Rwandan program, where conditions are drier, it has proved useful to wait until cuttings are ready to be transplanted.

The selection fields are established by the hill-plot technique. Each seedling is planted at the top of a 0.5–1.0 m^2 hill. After 6–8 weeks, three cuttings are taken from each seedling and planted at the base of the hill. The family structure is maintained, and, when the size of the families is adequate, the whole selection field is divided into four randomized blocks. At about 3–4 months, the whole selection field is screened for possible virus and anthracnose symptoms. The selection fields are harvested at 6 months. In future, some selection fields will be harvested at 4–5 months so that early-maturing material is identified. The selection intensity is 10%, and the grid system of Gardner (1961) is used — the 10 best hills from each successive 100 hills are selected.

The traits recorded for each seedling (hill) are: virus incidence; anthracnose symptoms; weevil damage (on a scale from 0 to 5); flesh colour; weight of vines; tuber yield; and number of tubers. Yield is the principal selection criterion, although all clones showing virus symptoms or serious weevil damage are culled. More than 6000 seedlings have been screened, including 1000 at high elevation (at ISAR-Rwerere) and 500 under subhumid conditions (at ISAR-Karama).

Orange-fleshed seedlings are generally low yielding and more susceptible to weevils. Selection fields arising from IITA seed material have more dropouts because of poor adaptation to Rwanda; therefore, selection intensity was increased to 2–3%. Moreover, IITA progeny show high virus susceptibility under Rwandan conditions, in spite of the fact that most IITA parents originated from a strong selection against virus (Hahn, elsewhere in this volume). This finding strongly suggests that viruses or their vectors are different in Rwanda.

Preliminary yield trials (PYTs) consist usually of four to five randomized blocks in which the selected clones are compared. Each clone is planted on a 1-m^2 hill plot, three cuttings at the top and five cuttings at the base. No border rows are planted between the different hill plots because in Rwanda sweet potato is always grown in varietal mixtures of at least four to five clones (Camerman and van Bellinghen 1973). Hence, one ought to select clones that exhibit high-yielding capacity under severe intergenotypic competition (Schutz et al. 1968).

A selection intensity of 10% is adopted at harvest. The selection criteria are the same as in the selection fields, although root shape is also noted. A sample of each clone is taken for taste rating and for analysis of dry-matter content, protein, saccharose, and starch. The best 10% of the clones are immediately multiplied. After multiplication, these clones are tested in multilocality yield trials (MLYTs) throughout Rwanda, during both the November and the March planting seasons. Some test sites are in swamps, and the Rubona planting site is duplicated with high inputs of fertilizer. Because of the scarcity of large testing sites in Rwanda and because of the extremely heterogeneous soil conditions, block sizes have been minimized so that the differences between them will be as few as possible. Plot sizes also have been reduced to four 1-m^2 hills/plot in a square configuration. Again, there are no border rows between plots. The number of main blocks has also been reduced to two, and two tiny blocks have been joined to the main experiment, and, within them, the clones are evaluated for their ability to compete with other genotypes. Each clone is allotted a mini-plot (80 cm × 30 cm) on which two cuttings are planted without a border row.

The same selection criteria as in the earlier selection stages (SFs, PYTs) are applied. Moreover, the clones are rated for environmental adaptation and stability, drought and cold tolerance. About 15 localities have been tested, including Kisozi (Institut des sciences agronomiques du Burundi, ISABU) and Mulungu (INERA, Zaire). In future, it is hoped to include Uganda and Tanzania.

The best 20% of the clones over all sites are selected. In addition, the best 20% of the clones at high elevations are added to the specific high-elevation subpopulation. Similarly, the best 20% drought-resistant clones are added to the drought-resistance sub-

Table 1. Sweet-potato-selection procedure used at ISAR.

Timing (year, month)	Selection stage	Clone population	Experimental layout	Selection criteria	Comments
1,8	Germination	12500	Incubator at 30°C	Germination	Germination is ~80%
1,8	Nursery	10000	30 m × 1.2 m beds, 10 cm × 5 cm spacing		
1,11	Selection field (SF)	10000	4 randomized blocks, 0.5–1.0 m^2 hills, 1 seedling + 3 cuttings/hill, grid system	Resistance to virus, anthracnose, weevil (scoring 0–5); yield of tubers and vines; number of tubers; flesh colour	Selection intensity is 10% within each grid of 100 hills
1,5	Preliminary yield trial (PYT) in swamp (dry season)	1000	4 randomized blocks, 1-m^2 hill plots, 8 cuttings/ hill (32 cuttings/clone), <250 entries/PYT	Same as for SF; also skin colour; taste; dry matter; starch; protein; saccharose	Selection intensity is 10%
2,11	Multiplication	100	na[b]	Virus resistance	Selection intensity is ~80%
2,3	Multilocation yield trial (MLYT) 1st season	80	~15 sites, 2 randomized blocks, 4 × 1-m^2 hills/ plot, 2 tiny randomized blocks for intergenotypic competition	Same as for SF; also environmental adaptation; stability	Selection intensity is 10% for each agroecological region
3,11	MLYT, 2nd season	80	Same as for MLYT, 1st season	Same as for MLYT, 1st season	Same as for MLYT, 1st season
3,3	Cultural practices	8[a]	As appropriate	Farmers' criteria	3–4 best clones/region are retained for release; not more than 10 clones for the whole of Rwanda
3,3	Intercropping	8[a]	As appropriate	Farmers' criteria	
3,3	Taste panel	8[a]	As appropriate	Farmers' criteria	
3,3	Multiplication	8[a]	As appropriate	Farmers' criteria	
3,3	Farmers' trials (FTs), 1st season	8[a]	24 farmers (2/region), 2 randomized blocks, 10–12 m^2 plots	Farmers' criteria	
4,11	FTs, 2nd season	8[a]	Same as for FTs, 1st season	Farmers' criteria	
4,3	Release	3–4	na	na	Release is through extension services and multiplication centres

[a] Along with eight selections, two local varieties are tested in each region for comparison.
[b] na = not applicable, not a trial.

population. In future, selection intensity will be increased to 10%.

The best MLYT clones are submitted to preextension trials (PETs), including trials on cultural practices and in farmers' fields done simultaneously along with taste-panel testing and large-scale multiplication.

In the trials on cultural practices, different levels of fertilizer inputs and planting density are combined, the goal being to determine the optimum agronomic environment for the best clones. The effects of interspecific competition are studied in intercropping trials (ICTs) with the best clones.

Farmers' trials (FTs) include the same clones, which are planted in farmers' fields all over Rwanda. A taste panel (TP) is organized with 20 farmers to rate the best clones for taste, and the clones are multiplied on a large scale under phytosanitary control.

These preextension trials enable the choice of three or four very good clones to be released for each of the 12 agroecological regions of Rwanda. Because there is some overlap in the conditions in the 12 regions (Delepierre 1974) and some environmental stability in the new selections, it is hoped that the total number of clones for release and multiplication can be reduced to about 10.

Recent experimental results

In a preliminary yield trial with 20 clones conducted in 1979, a correlation analysis was performed on eight traits. Tuber yield was positively correlated with average tuber weight ($r = 0.88^{**}$) as well as with number of tubers per unit of surface ($r = 0.78^{**}$). This finding confirms earlier work by Jones (1970) and Mahungu (1979) indicating that multiple tuberization increases yield.

Top growth is a component of yield quality, as it is positively correlated with starch content ($r = 0.49^{**}$) and negatively correlated with water content ($r = -0.46^{*}$). This explains why Rwandan farmers prefer starchy varieties, which are associated with low-water content, i.e., better storage and higher nutritive value.

In a multilocality yield trial in 1979 with 19 clones on seven sites throughout Rwanda, one clone, Rusenya, showed clear superiority over 18 others. It yielded, on average, 8.1×10^3 kg/ha over a 6-month growing period, compared with a grand mean of 3.9×10^3 kg/ha. Two other clones showed good yield potential: Caroline Lee (5.7×10^3 kg/ha) and Bukarasa (5.2×10^3 kg/ha).

The genotype × environment interactions, which were highly significant, were plotted graphically, and a cluster analysis was performed. It was possible to identify groups of varieties adapted to ideal conditions (e.g., Bukarasa), cold conditions (e.g., Rusenya, Divir 1318, D.Virovsky 16, var. Péruvienne), and dry conditions (e.g., Nyirakayenzi, Nyiramujuna), which were then added to the appropriate breeding subpopulations.

Future breeding strategy

Reciprocal recurrent selection was first proposed by Comstock et al. (1949) as a means of selecting simultaneously for both general and specific combining ability. The method was modified by Eberhart et al. (1967) into a "comprehensive breeding system" whereby recurrent selection for favourable characters is performed within each of two or more populations, and the two or several improved populations are intercrossed. The usefulness of population breeding has been adequately demonstrated by Suneson (1956).

Jones (1965) proposed that a random-mating population be maintained and subpopulations selected toward different goals. He later reported that after only six random-mating generations some desirable agroeconomic characters within the population had decreased — vine diameter, root weight, skin smoothness, resistance to *Fusarium* — and vine length had increased (Jones 1972). However, a higher degree of resistance against Coleoptera was found (Cuthbert and Jones 1972).

At IITA, short-range (weevil and virus resistance) and long-range (large genetic variability) population improvement is undertaken simultaneously.

Multireciprocal, recurrent population improvement

A central, breeding-population pool will be constituted for trials on yield potential at mid-elevation conditions (1500–1900 m) where most of the sweet-potato cropping is done. Concurrently, a set of interdependent-breeding subpopulations will be constituted, each subpopulation corresponding to a particular breeding goal (Jones 1965), e.g., resistance to cold, drought, hydromorphic conditions, virus, and weevil; early maturity; and content of saccharose, carotene, protein, starch. There will be continuous multireciprocal exchange of suitable genotypes between the central breeding pool and each of the respective subpopulations, as well as between the subpopulations themselves. Hence, in each population the frequency of the desired genes is likely to increase steadily (Eberhart et al. 1967).

Each population will comprise:

- A recurrent-breeding population of parents known to be outstanding with regard to a particular trait; this recurrent-breeding population (subcollection) will be enriched gradually with proven material selected from yield trials;
- A fluctuating breeding population constituted by the clones, selected in the preliminary yield trials; and
- A random-mating population of 1000–2000 seedlings — the amount depending on land availability.

However, to reduce genetic drift toward undesirable characters (Jones 1972), there will be a seasonal 10% enrichment of the random-mating population by the MLYT clones retained for farmers' trials. Hence, each season a random-mating population will comprise about 90% seedlings and 10% clones.

Reciprocity between the central breeding pool and each of the subpopulations is easy to achieve: in each of the recurrent-breeding populations (subcollections), clones known for their high-yielding capacity will be interplanted for pollination of the subcollection; seed harvested on these pollinators will be taken back to the selection fields of the central breeding population. Moreover, some selections from the central PYT will be diverted to the PYT stage of specific populations, and vice versa. Similarly, PYT selections with specific characters will be interchanged between the corresponding subpopulations.

Selection procedure within the central breeding population

Within the central breeding population, the improvement program will incorporate several steps in addition to those in the overall improvement program (Table 1):

- Seed will be harvested from the whole breeding population; two or three seeds will be taken for each plant in the random-mating population (RMBP) and will be bulked together; all available seed for the fluctuating breeding population (FBP) will be harvested and bulked together; the recurrent breeding population (RBP) seed will be harvested, and the progeny structure, whether open-pollinated or control-pollinated, will be maintained according to an appropriate mating design.
- The clones selected under PYT will be advanced into MLYT; some PYT selections will be added to the PYT stage of specific subpopulations; moreover, all MLYT clones will be put into a crossing block and will constitute the next fluctuating breeding population; in addition, one cutting of each MLYT clone will enrich the random-mating breeding population.
- The clones selected after MLYT will not only be advanced to PET but also be added to the recurrent breeding population, and some of these PET clones may be added to specific RBPs.
- Flowering-induction techniques will be used wherever breeding populations (RMBP, FBP, RBP) are grown.
- At harvest, the RMBP will be considered just as another selection field.
- The RMBP will be initiated by the collection of 10–20 seeds from each of 50–100 parents.
- New clonal introductions will be added straight to the PYT stage, whereas seed introductions will first be screened into SF.

Selection procedure within each specific subpopulation

The selection procedure for each specific subpopulation is as for the central population. However, a few particular measures are to be taken:

- The selections arising from PYT will join the PYT stage of the central population.
- The clones arising from given subpopulations and reaching the FT level will enrich the relevant specific subcollections (RBP).
- In each of the RBPs (subcollections), the central RBP (aimed at yield potential) will be interplanted for interpollination.

I wish to acknowledge the Belgian Overseas Aid Program (Administration générale de coopération au développement, Brussels) for financing sweetpotato research in Rwanda.

Sweet-potato improvement in Cameroon

H.J. Pfeiffer[1]

The major factors limiting the extension of sweet-potato culture in central–south Cameroon are: low and irregular yields; susceptibility to weevil and virus attack; and difficult integration into traditional farming systems. Thus, these factors are the focus of sweet-potato improvement efforts by the Cameroon National Root Crops Improvement Program (CNRCIP), which has undertaken collection of local clones, yield trials with local and imported clones, and selection for resistance to virus and weevil attack. Results to date indicate that the local clone from Nkolbisson performs better over one 6-month season than over two 4-month seasons and that seed introductions from the International Institute of Tropical Agriculture (IITA) may permit sweet potatoes to be grown over two 4-month cycles, with an accumulated yield of 30–35 t/ha. Covering range and speed are being investigated, as they are especially important during the 2nd season. Tests are also being conducted to determine the acceptability of new clones in terms of taste, texture, and sugar content.

Les principaux obstacles à l'extension de la culture de la patate douce dans le centre et le sud du Cameroun sont des rendements faibles et irréguliers, la sensibilité aux attaques des charançons et des virus et les difficultés d'intégration dans les systèmes agricoles traditionnels. Ces problèmes sont l'objet des travaux d'amélioration de la patate douce du DNRCIP, qui procède à la collection des clones locaux, à des essais de rendement avec des clones locaux et importés et à la sélection en fonction de la résistance aux charançons et aux viroses. Les résultats obtenus jusqu'ici indiquent que le clone local de Nkolbisson a un meilleur rendement avec un cycle de six mois qu'avec un cycle de quatre mois, et que l'introduction de semences de l'Institut international d'agriculture tropicale (IITA) pourrait permettre deux cycles de culture de quatre mois et un rendement global de 30 à 35 t/ha. On étudie la superficie et le rythme de production, très importants lors de la deuxième saison. On fait aussi des essais pour déterminer si le goût, la texture et la teneur en sucre des nouveaux clones sont acceptables.

Sweet potatoes were probably introduced to Cameroon from tropical America about the end of the 19th century. Their cultivation was only slowly adopted because of the sweet taste of the tubers and the fact that they were difficult to integrate into the traditional farming system. However, they have become popular in the high-altitude areas (North-West and West provinces), which produce three-quarters of the country's total production (Table 1).

Sweet potatoes represent only a small part of the total root and tuber production and are, therefore, often considered a minor crop. The yields calculated from the cultivated areas in 1978–79 appear to be very low; however, because sweet potatoes are mostly produced in a mixed-cropping system, the estimated area, in fact, represents all cultivated areas where sweet potatoes are found.

It has been estimated that sweet potato has a potential yield of 7.6×10^3 kg/ha in Cameroon over a normal cropping period of 5–6 months (Project PNUD/FAO). However, potato cultivation is poorly developed. It was against this background that a national root crops improvement program was created within the IRA (Institut de la recherche agronomique) with the technical supervision of IITA and the financial support of the International Development Research Centre (IDRC, Canada) and the Administration générale de la coopération au développement (AGCD, Belgium).

[1] Cameroon National Root Crops Improvement Program (CNRCIP), Institut de la recherche agronomique, B.P. 2067, Yaounde, Cameroon.

Table 1. Production of sweet potato in Cameroon (1978–79).

	Cultivated area (ha)	Production (t)	Yield (kg/ha)	% of total root and tuber production
North	15575	8338	535	4.7
East	10681	479	45	0.4
Centre–South	30911	4124	133	1.2
Coast	9941	4087	411	1.9
Southwest	3139	807	257	0.4
West	53433	30176	565	6.3
Northwest	35573	19933	560	4.8

Present situation in central–south Cameroon

Sweet potato represents only 1.2% of the total root and tuber production in the centre–south (Table 1). In the forest area, sweet potatoes are generally cultivated around the houses on mounds and used to fill spaces in the cassava fields after the harvest of groundnuts and maize at the end of the 1st cropping season. In the savanna and forest–savanna transition area, pure sweet-potato stands in small fields (100–500 m^2) are commonly found. Some sweet potato is also grown on mounds around the houses.

Mounding is common. The mounds are normally 30–50 cm high and spaced about 1.5 m apart in each direction. Planting is done with vines, about 1 m long, that are shaped into a 15-cm diameter ring and planted in the top of the mound. Some farmers plant vines that are 30–40 cm long. Usually, two to four cuttings are planted per mound, and about half the length of the vine is buried.

Planting is generally done between mid-May and mid-July, and harvesting takes place after 5 months, although the time depends on family needs. Only one weeding is done, generally between the 1st and 2nd month. Around the houses, sweet potato is cultivated permanently, and this crop provides the planting material for other fields.

Only a limited number of clones are grown in central–south Cameroon, and two clones with a low-production potential and poor tolerance to weevil attack are the most common. Because of this very low variability at the regional level, a national local collection of 65 clones was gathered at Nkolbisson during the 1st season of 1979. Of these clones, 13 (PA series) came from another collection and had already been tested in high-altitude areas. The general performance of the clones was relatively poor, and the best clone in 1979 (Nyombe 5) yielded 8.6 t/ha over a growing period of 4 months at a density of 30 000 stalks/ha.

After some identical clones were regrouped, and others were lost because of a lack of multiplication material (heavy virus attack), 34 clones were retested in 1980 over a growing period of 5.5 months. The best clone (Caib 4) produced 15.7 t/ha (Table 2). Six other clones yielded between 10 t/ha and 15 t/ha, but 18 yielded less than 5 t/ha. Resistance to diseases and pests in such a collection, coming from different ecological zones, was very difficult to evaluate, and no relation could be established between the results of 1979 and those of 1980. Generally, most clones were more sensitive to virus attack than the local clone (Nkolbisson 1). Resistance of the local varieties to the

Table 2. Characteristics of the seven best local clones in 1980 (Nkolbisson) after 5.5 months (30 000 stalks/ha).

	Yield (t/ha)	Form	Colour	
			Skin	Flesh
Caib 4	15.7	Oval	Red	Yellow-violet
Pa 2	15.0	Long	White	White
Santchou 1	14.9	Long	White	White
Caib 1	12.8	Oval	White	White
Pa 8	10.5	Oval	Red	Cream-yellow
Nkongsamba 1	10.1	Long	White	White
Ciam 5	10.0	Long	White	White
Local 1	5.0	Long	White	White

sweet-potato weevil, especially in high-incidence areas, seemed to be low compared with some of the new introduced clones.

In summary, the major constraints limiting the extension of sweet-potato culture in centre–south Cameroon are: the low and irregular yields compared with the amount of work needed and with the performance of other crops (cassava, cocoyam); the susceptibility of the local clones to the weevil (*Cylas puncticollis*) and to virus attack; the difficulty of integrating sweet potatoes into the traditional farming systems; the necessity to maintain live material during the dry season, a constraint strongly limiting the potential area for cultivation; the relatively low prices compared with yam as well as the difficulty of storage at the farm level; and the low range of acceptable quality.

Initial research results

The first cultivars imported into Cameroon in the early 1970s came from the USA, Taiwan, and Nigeria (IITA). From this material, only the cultivars TIB 1 and TIB 2 (IITA) showed high performance in the high-altitude areas. These two clones were distributed to farmers and were tested in 1979 (2 seasons) and 1980 (1st season) in centre–south Cameroon (Table 3). In these trials, the imported clones outyielded the local ones except during the 2nd season at Kribi. Acceptability by the local people did not present any problems, and the two clones seemed to have a wide ecological adaptability. Their yields at Bambui (1600 m altitude) were 20.5 t/ha and 16.0 t/ha, respectively, over 5 months. However, even though their yields were 2–3 times those of local clones, their general performance and their resistance to weevils were not satisfactory. For that reason, new seed introductions from IITA elite parents were obtained at the end of 1978 and in mid-1979. The first introduction included 114 clones from 9 families (Table 4); the second introduction included 647 clones from 12 families (Table 5).

Since the 2nd season of 1979, a standardized selection program has been used. Approximately 12 500 seedlings have been raised in the nursery; 5500 clones from 9 IITA and 112 Nkolbisson families were planted in December 1979 for preliminary observation, and 285 were selected for the first clonal evaluation (1st season of 1980). Some of these clones proved very promising.

Evaluation of results

Yield and adaptability

The mean yields of Nkolbisson 1, grown over different periods at three locations, were: at Mbankomo, 1st season (4 months) 8.9 t/ha; 2nd season (4 months) 3.6 t/ha; one 6-month season 12.9 t/ha; at Nkolbisson, 1st season (4 months) 5.8 t/ha; 2nd season (4 months) 1.95 t/ha; one 6-month season 9.8 t/ha; and, at Nkometou, 1st season (4 months) 6.5 t/ha; one 6-month season 10.3 t/ha.

These results indicate that growing the local clone (Nkolbisson 1) for one cycle of 6 months results in yields that are identical or even superior to those obtained over 2 seasons of 4 months each and requires less work. However, it should be possible to produce greater yields from two consecutive crops than from one, and centre–south Cameroon has a bimodal climate — a season from April to August and another from September to December. Growing two crops would make it

Table 3. Yield (t/ha) of TIB 1, TIB 2, and local clone over a growing period of 4 months.

Locality (season)	Rainfall (mm)	TIB 1	TIB 2	Local
Nkolbisson (1)	1700	12.3	10.1	3.9
Mbankomo (2)	1600	7.6	7.3	3.5
Niga (1)	1500	14.8	–	4.5
Nkometou (1)	1500	11.2	16.5	6.5
Essasok (1)	1500	32.0	20.0	3.4
Hondole (1)	1700	17.8	–	5.3
Kribi (1)	3000	6.8	7.6	1.1
Kribi (2)	3000	12.3	2.1	(1) 3.8
				(2) 14.7

Table 4. Performance of 16 best clones (after 3 selection seasons)[a] from 114 clones (IITA material).

Clone	Mean yield (t/ha)	Virus susceptibility[b]		Weevil attack (V/T)[c]		Dry matter (%)	
		1979	1980	1979	1980	1979	1980
8 4	16.2	0	0.35	1.2/1.5	2/0	30.8	29.1
8 1	16.3	0.5	0.75	0.5/1.5	1/0	38.3	27.0
2328 38	15.9	0.5	0.9	1.5/3	2/0	35.8	33.0
3017 1	15.3	0.25	0.75	1.5/2.5	1.2/0	33.1	31.7
3290 4	14.7	0.5	0	1/2	1.5/0	39.2	–
2328 10	14.4	0.5	2.0	1.0/1	1.7/0	36.5	32.4
8 20	14.2	0	1.8	1.2/1.5	2.2/0	36.9	31
2330 2	13.3	0	0.7	1.5/1	1.5/0	32.9	23
8 24	13.0	0	0.1	1.7/2	1/0	39.8	39
3017 7	13.0	0.5	0.75	1/1	2.5/0.7	34.3	30
8 22	13.0	0.5	0.7	1/1	2/1	33.4	30.5
8 13	12.7	0	0.6	1/1	1.5/0	41.2	31.0
8 6	12.7	0.5	0	1.5/1	2.2/0.5	36.9	27.9
8 35	12.1	0	0.5	0.5/1	1.5/0	31.4	30.7
3247 1	11.9	0	1.0	1.5/2	2/0	33.8	–
8 21	11.7	0	0.1	0.9/1	1.5/0	42.0	33.8
Local	4.4	0.1	0.5	1.4/2.8	2/0.2	39.6	34.3

[a]Based on 1st season at Nkolbisson in 1979; 2nd season at Mbankomo in 1979; and 1st season at Niga, Nkometou, and Essasok in 1980.

[b]Virus susceptibility estimated by the mean frequency of the symptoms: 0 = 0%; 0.5 = 10%; 1 = 25% of the plants.

[c]Weevil attack on vines and tubers (V/T): on vines: 0 = no attack; 1 = low attack; 2 = medium attack; 3 = heavy attack; on tubers: 0 = no symptoms; 1 = light symptoms of yellowing, vein clearing, or mottling without significant leaf-size reduction; 2 = more severe symptoms with significant leaf-size reduction; and 3 = very severe symptoms with stunting of plants.

Table 5. Characteristics of the 15 best clones (after 2 selection seasons) from 647 clones (IITA material).

Clone	Mean yield (t/ha)[a]	Virus[b]		Weevil (V/T)[b]		Dry matter (%)[d]	Tuber colour[e]	
		Nkolbisson[c]	Essasok[d]	Nkolbisson[c]	Essasok[d]		Interior	Exterior
2 26	19.2	0	0.5	1/1	1/0	32.1	Y	Ro
2 21	18.4	0	1.5	1.5/0.5	1.5/0	34.6	Y	R
11 12	16.5	0	0	1.5/1	1/1	29.0	C-O	Co
2 24	15.1	0	1.0	2/2	2.7/0	34.3	Y	Ro
330 3	20.2	0	0	1/2	1/0	–	B	R
11 10	14.4	0	0	1/1	2/0	28.0	Y	Co
270 34	16.3	0.1	0	1.5/2	1.5/2	33.2	C	P
11 45	12 7	0	1.0	2/2	1.5/0	29.2	C-Vi	R
11 62	12.0	0.5	1.5	1.5/2	1.5/0.5	–	Do	R
11 19	11.6	0	0.5	1.5/2	1.5/0	–	C	Co
2 9	9.6	0	0	1.5/2	–	–	O	R
8 43	10.3	0	1	1/0.5	1/0	28.8	C	B
270 54	7.3	0.5	0	1/2	1.5/1	–	O-Vi	P
11 8	11.3	0	0.5	1.5/1	2.5/0	29.0	Y-O	Co
125 13	6.6	0	2.5	1/2	2/0	32.6	C	R
Local	4.0	0	1.0	2.83	2/0.2	34.3	Y-C	B

[a]Based on 2nd season at Nkolbisson in 1979 and 1st season at Essasok and Nkometou in 1980.

[b]See Table 4 for codes.

[c]2nd season of 1979.

[d]1st season of 1980.

[e]Colour code: Y = yellow; C = cream; O = orange; B = white; R = red; Ro = rose; Vi = violet; Co = copper; P = purple; Do = deep orange.

easier to integrate sweet potatoes into the local cropping systems. For these reasons, selection should favour clones with early maturity (4-month growth cycle). Because yields in the 2nd season are generally low compared with those in the 1st season, cultivars with good yield potential over the 2 seasons should be selected. The cultivation of two different clones in the 2 seasons is not practical at present for farmers in the area.

The principal ecological zones in centre–south Cameroon, characterized by their natural vegetation, are savanna, transition, and forest. The adaptability of the different cultivars is under study. However, from 761 clones tested at Nyombe (80 m altitude and 2000 mm rainfall) and at Nkolbisson in 1979, only 5 clones were selected at each site after 2 seasons.

Resistance to viruses and weevils

During the 2 seasons of 1979, few virus symptoms were observed in the selection plots (except for family 5270 where 57% of the clones showed symptoms). However, during the dry season (1979–80), the multiplication of the selected material was done with tubers (for line-purification reasons), and 36% of the clones showed virus infection. Of these, 95% reexpressed symptoms during the 1st season of 1980 in at least four locales despite the fact that the trials were established from vines that were apparently clean.

Generally, a clone that shows symptoms in one locality will do so in others, even if the frequency and the severity of attack differ quite widely. High frequency and severity of virus symptoms were observed in the trials at Lolodorf (400 m altitude, 2000 mm mean rainfall), but sweet potatoes are so rarely cultivated in the area that it was impossible to find a local clone for the trial. Some clones have not yet exhibited any symptoms of virus attack, and others have only shown a low rate of infection.

The sweet-potato weevil (*C. puncticollis*) is the worst enemy of this crop in centre–south Cameroon. The damage caused by this insect to vines and tubers strongly limits sweet-potato production and commercialization. The damage on the vine is similar for the 2 growing seasons (Tables 4 and 5); however, the damage caused to the tubers is much more severe during the 2nd cropping season. During the 1st season, attacks are rare and are mostly limited to the tuber neck. The major reason for heavier attacks during the 2nd season is that there is almost no rainfall during the last 5–8 weeks, and many cracks appear in the dry soil. The size of the tubers at this time also enhances this phenomenon, allowing the weevils to reach the tubers more easily. Moreover, the temperature and humidity favour development of the weevil population.

No resistance to weevil was identified, but relatively important differences in the behaviours of the clones (grouped by family) were observed (Table 6). The results were not consistent; for example, no relationship could be established between the damage seen on vines and that observed on tubers in a trial at Mbankomo during the 2nd season of 1979; the clones of families TIS 2328 and TIS 2532 were

Table 6. Intensity (%) of weevil attack for different sweet-potato families evaluated during 2nd season, 1979, at Nkolbisson.

Family	Number of clones	Intensity of attack[a]				Selected clones		
		Low (0–2)	Medium (3–4)	Heavy (5–6)	Very heavy (7 +)	Number	Class: Low	Class: Medium
TIS 3030	58	21	26	19	34	0	–	–
TIS 5270	61	10	20	22	48	3	3	0
TIS 3017	41	41	5	17	37	2	2	0
TIS 3247	43	2	12	30	56	1	1	0
TIS 2532	46	13	20	15	52	1	1	0
TIS 2330	30	37	27	23	13	2	2	0
TIS 5003	22	9	14	36	41	1	1	0
TIS 3290	17	0	12	6	82	0	–	–
TIS 5125	56	9	36	16	39	5	5	0
TIB 11	82	12	13	25	50	8	7	1
TIB 2	59	20	24	39	17	5	3	2
TIB 8	79	8	18	22	52	3	2	1

[a]Local clone: mean attack = 2.83.

heavily attacked on the vines, and TIB 8 showed variable behaviour. The clones of family TIB 8 were attacked at a very low level on their tubers at Mbankomo, in contrast with the findings at Nkolbisson, but the two groups of clones originated from different introductions. At Nkolbisson, sweet potatoes were cultivated for 2 consecutive seasons on the same land — a practice that encourages a buildup of the weevil population.

Covering range and speed

Because of the heavy rainfall and the erosion-sensitive tropical soils, a quick and adequate soil coverage, especially at the beginning of the 2nd rainy season, is required. At the density of 30 000 stalks/ha, the best clones cover 50% of the land after 4 weeks and completely cover the surface after 7–8 weeks. Cover density depends on leaf area and orientation as well as the number of leaves per unit length of the vine.

The selected clones are great improvements over the local clone, which generally completely covers the soil after 10–12 weeks and has a low cover density (small leaves and few leaves per unit length of vine).

Sweet-potato quality

A sweet potato not only must be excellent from an agronomic point of view but also must be acceptable to the consumer in terms of taste, colour, and smell and be adapted to the local cooking practices. A preliminary acceptability test involving 20 people was carried out in January 1980 on the harvest of the 2nd season of 1979. Everyone agreed on the high quality of the local clone (Nkolbisson 1), and the other local clones (CIB 23 and CIB 46) were well liked. Flesh colour was important: deep orange and violet (or purple) were rejected, whereas clones with white or yellow flesh were generally well liked. Dry-matter content also seemed to be important: high values correspond to good or very good taste, whereas values lower than 34–35% resulted in bad to very bad scores in the taste tests. Sweetness and texture were also noted, but the terms of the evaluation must be improved.

Prospects

If sweet potatoes in central–south Cameroon are to be made more attractive to farmers and consumers, the improvement of genetic variability by introductions and crossings must be a high priority.

Sweet potato easily sets seeds under the conditions at Nkolbisson, and more than 22 000 seeds from 305 families were collected in open-pollination plots during the 2nd season of 1979. Isolated blocks including the best-yielding clones and those that were most tolerant to weevils were initiated during the 2nd season of 1980 in two different locations for seed production. A selection program was also initiated in the 1st season of 1980. From the point of view of yield, optimum production over 2 growing seasons will be the major goal. Work is also required on the integration of the new material into the traditional mixed-cropping system, and selection for resistance to diseases and pests is essential. For viruses, the tuber-planting method for more rapid detection of virus-sensitive clones will be retested during the dry season. Virus scoring in the field will be done more precisely: attack severity and frequency will be combined.

The vine reaction (swelling) to weevil attack is variable and seems to be quite specific to each clone. This, along with the section of the vine occupied by the larvae, seems to determine the final yield, especially when the plant is attacked early. A weevil-scoring key has, therefore, been developed as a means to evaluate and document the influence of vine attack on yield and to help specify the reaction of each clone.

In terms of product quality, acceptability tests will be conducted from the first yield trial. In addition, research will be continued in an effort to specify the physically measurable elements (colour, dry-matter content, etc.) that are deemed important by the local population.

The storability of different clones must also be evaluated under farm conditions. And, finally, the use of sweet-potato leaves as a vegetable should be promoted, and clones adapted to leaf production should be given special attention.

Strategy for developing a national potato program for Rwanda

P. Vander Zaag[1]

The potato, which was introduced into Rwanda about 1900, has become a major food crop in only the past 35 years. Production per unit of land is low (6.8 t/ha) because of a lack of quality seed potatoes of adapted varieties; infections from *Phytophthora infestans* and *Pseudomonas solanacearum*; poor agronomic practices; poor preparation of seed potatoes; and an acute lack of technical knowledge among extension personnel and farmers. The Programme national pour l'amélioration de la pomme de terre (PNAP) was established to help remove these constraints to production. A research program with a practical approach to solving problems has been established along with a strong seed production, training, and extension program. The results after 1 year have been encouraging. On one hand, farmers are accepting seed production and storage techniques and see the utility of using fungicides and improved varieties in controlling *P. infestans*, and, on the other, PNAP scientists have gained valuable information from rural contacts.

La pomme de terre, introduite au Rwanda au début du siècle, ne fait l'objet d'une culture vivrière importante que depuis trente-cinq ans. La production par unité de terrain est faible (6,8 t/ha) vu le manque de tubercules de qualité des variétés adaptées, les infections dues au *Phytophthora infestans* et au *Pseudomonas solanacearum*, les façons culturales inadéquates, la mauvaise préparation des fragments de plantation, et un manque critique de connaissances techniques parmi le personnel local et les cultivateurs. Le Programme national pour l'amélioration de la pomme de terre (PNAP) a été mis sur pied pour pallier à ces obstacles. On a mis en oeuvre un programme de recherches visant à trouver des solutions pratiques aux problèmes ainsi qu'un vaste programme de production, de formation et de vulgarisation. Après un an, les résultats sont encourageants. D'une part, les cultivateurs acceptent les techniques de production et d'entreposage des pommes de terre et voient l'utilité des fongicides et des variétés améliorées pour lutter contre le *P. infestans* ; d'autre part, les chercheurs du PNAP ont obtenu de précieux renseignements de leurs collaborateurs ruraux.

Legend has it that German colonists at about the turn of the century unwittingly planted potatoes throughout Rwanda as they traveled with sacks of potatoes thrown over their horses' backs, an occasional tuber falling and becoming established. At night, the travelers would set up camp and cook their meal, discarding damaged tubers and sprouts, which under favourable conditions would reproduce themselves.

In 1904, Belgian missionaries grew potatoes for the first time in their garden near Ruhengeri. European varieties and cultivation techniques were spread quickly throughout the country by the missionaries as they established new missions. They also familiarized the Rwandaise with the previously unknown vegetable. However, the potato was not rapidly accepted into the local diet primarily because many held the superstition that eating potatoes along with drinking cow's milk would result in the death of the cow. The local chiefs realized, after watching the Europeans, that the potato could be eaten with no risk, and potato consumption increased markedly following the famines of 1928–29 and 1944–45.

In 1930, the first European variety was introduced and tested at the request of the gov-

[1] CIP/Programme national de l'amélioration de la pomme de terre (PNAP), Institut des sciences agronomiques du Rwanda, B.P. 73 Ruhengeri, Rwanda.

ernor of Rwanda-Urundi. Mr Dejeune, a Belgian agronomist, introduced more European varieties after 1936, which were followed by introductions by the Institut national d'études agronomiques au Congo (1951–63) and the Institut des sciences agronomiques du Rwanda (1963–79). Varieties introduced originated from Belgium, Holland, Germany, Australia, Mexico, Kenya, and Uganda.

According to the Ministry of Agriculture (in 1978), potato production has increased steadily, reaching 2×10^5 t/year from an area 3.2×10^4 ha. In the major potato-growing areas, per-person consumption is nearly 1 kg/day at present (S. Poats, unpublished data).

Present situation

The potato grows in every prefecture of Rwanda, with 90% of the production in the prefectures along the Nile–Zaire divide at elevations greater than 1800 m (Byumba, Ruhengeri, Gisenyi, Kibuye, and Gikongoro). At these high altitudes, the temperatures are moderate and rainfall is 1200–2000 mm, which makes these areas unsuitable for crops such as sweet potatoes, cassava, and sorghum. In the past 10 years, the volcanoes along the northern borders of Rwanda have been partially deforested, and potato production has gained primary importance on these rich volcanic soils at elevations of 1800–2800 m.

In mid-1979, upon the inception of the Programme national pour l'amélioration de la pomme de terre (PNAP), an in-depth evaluation was carried out to determine research priorities. Visits were made to the major potato-growing areas; discussions were held with the local government leaders, extension workers, and farmers in their fields. During these discussions and observations, problems that were most frequently mentioned were:

- The lack of seed potatoes of adapted varieties. Farmers utilize the small potatoes from the previous harvest for seed. Often these small tubers are from virus-infected plants or from bacterial-wilt-infected plants. Other sources provide seed for less than 0.25% of the total area planted each year.
- Late blight (*Phytophthora infestans*) and bacterial wilt (*Pseudomonas solanacearum*). The heavy rains along with ideal temperatures allow both diseases to damage the potato crop severely. The European varieties generally have little resistance to the two diseases. Although, recently, several varieties from Mexico (Sangema, Atsimba, and Montsama) have shown some resistance to late blight, farmers had not received these varieties in quantity as of 1979.
- Low yields in farmers' fields because of unsuitable practices and mixed-cropping patterns. The question is whether or not the potato varieties are at optimum total production per unit of land.
- Storage techniques. Seed potatoes simply originate from the potatoes stored for consumption. No effort is being made to store potatoes for seed separately or under conditions conducive for healthy sprout development.
- The lack of technical knowledge about potato production. Training of extension workers and farmers must receive priority if increased potato production per unit of land is to be attained.

Priority research program

A program of potato research having six major thrusts was established in October 1979. These thrusts are:

- Introduction and selection of new genetic material. Introductions have been broadened to include genetic material not only from Europe, Uganda, and Mexico, but also from Kenya and the Centro Internacional de la Papa (CIP), Lima, Peru. This material will be tested for resistance to late blight and bacterial wilt as well as production and quality factors. In the final stages (4th season after introduction), multiplication will commence so that a sufficient quantity of seed will be available for release. Efforts will be made to use some rapid multiplication techniques (sprouts, stems, or bud cuttings) for promising material.
- Agronomy and physiology. Experiments to test the effects of spacing, seed size, seed rate, hill size, desprouting, age of seed, rate of tuberization and development all will receive initial attention. Long-term studies will focus on fertility requirements and maintenance and cropping systems with the potato. The use of botanical seed, transplanting, and other agronomic practices will be studied in detail in the hope of transferring the technology to farmers.

• Pathology. Emphasis is being placed on the control of late blight by management practices such as regulating date of planting, using appropriate fungicides, regulating the frequency of chemical application, and determining varietal resistance. Long-term emphasis will be on developing and transferring the appropriate technology to farmers and on helping them to obtain the necessary inputs (sprayer, fungicide). Bacterial-wilt control through management practices is also being investigated. The effects of removal of infected plants and the soil from around the plant, different crop rotations, volunteer plants, alternative hosts, and races of bacterium present are being monitored. The long-range goal is to obtain varieties with blight resistance.

• Quality seed potatoes produced in quantity. A 40-ha seed farm at 2300 m elevation is being utilized to produce 300 t of improved seed (20 ha/year) for distribution to multipliers of seed potatoes who in turn distribute their seed potatoes to potato growers. It is hoped that, in this way, 2000 t of seed potatoes of good quality will reach the farmer; that the recipients will sell the seeds to their neighbours; and that each year different farmers will have access to the limited supply of seed. It is planned that, once every 5–6 years, all farmers will obtain quality seed potatoes.

• Storage and handling. Efforts are to focus on low-cost structures designed for seed-potato storage for farmers, as individuals or as groups or cooperatives, and for communes, projects, and organizations who will multiply seed potatoes for distribution. Later, efforts will focus on improving the storage of ware potatoes and particularly address the question of storage by communes or cooperatives when the price is low. The farmers may not be able to afford these storages so some financial arrangement may be necessary — for example, an agency to purchase from the farmer at a reasonable price during the harvest seasons and to sell when the demand exceeds supply. PNAP will study and propose the design of the buildings, but the need exists for other sources to finance and to make the storages functional.

• Training and extension. A concerted, long-term effort will be given to the training of all extension workers who are involved with potato production in Rwanda. Courses will be held at PNAP, Ruhengeri, for 1 week, and short courses (1 day) will be held at commune headquarters for farmers and local leaders.

During the initial 2 years, demonstration plots will be established with the collaboration of extension workers so that PNAP staff can determine the utility of farmer methods and, further, determine the constraints for adopting improved technology. Through these demonstration plots, the research program will remain relevant, as the focus is on farmers and their problems of production. PNAP's objective is to do research that will help farmers and that will be adopted by them.

The demonstration plots will be a learning mechanism for PNAP, the extension worker, and the farmer. The priorities for these demonstration plots include the development and use of:

• Seed-production techniques that employ simple steps to select healthy plants for seed, to identify diseased plants (those with virus and bacterial wilt), and to control infections.

• Fungicidal sprays, resistant varieties, and schedules for planting that help control late blight.

• Improved agronomic practices and mixed-cropping techniques.

• Improved storage of seed potatoes.

After the first 2 years, the initiative for demonstration plots will be in the hands of extension personnel and agricultural projects. PNAP will provide technical assistance as needed and be present for large-group training sessions at the demonstration plots organized by extension workers.

The collection and dissemination of information on the potato and its production will be a continual priority of PNAP. The training and extension thrust will occupy between 20% and 30% of the time of each PNAP scientist.

Program functioning

National scientific staff in the program comprise three university graduates (Ing. Agron.) and two agricultural-diploma holders. CIP has stationed two scientists in Rwanda for a maximum 5 years to assist in establishing PNAP. Their time is also devoted to establishing a national program in Burundi and assisting Uganda and Zaire in their programs. The identification of individuals to fill support roles, such as secretary, bookkeeper, storekeeper, mechanic, driver, field observer, and labour captain, has been difficult; however, most positions have now been filled. Support staff are very important, as they are permanent and carry on the daily activities. Well-trained, self-directed support staff are essential.

Each thrust of research has a leader and a co-leader, who, together, must develop an annual program of activities in keeping with the overall goals of PNAP. The proposed activities are discussed by the PNAP scientists and approved or revised. Then, weekly meetings are held for discussions on progress, problems, and new ideas related to the thrust.

Each scientific staff member has some administrative responsibilities: one is responsible for the functioning of the seed farm (support staff included), one for the research station, and one for the office and lab (personnel and equipment).

With this division of administration, all scientists can spend 80% or more of their time on scientific responsibilities and no one person is totally lost in administration. Furthermore, this allows all to gain some experience in administration to prepare for what is often the inevitable — a scientist functioning solely as an administrator.

The facilities at Ruhengeri research station include four houses for the scientists; another building houses the program — eight offices, an adequate laboratory, a handling room for potatoes, a store room, and a dormitory with a capacity of 10–12 beds. A greenhouse provides the space for pot experiments. A screen house is to be used for multiplication of promising clones destined for the seed farm. About 2 ha of land is available for experimental work.

The Kinigi seed farm, which covers 40 ha, includes a house along with two apartments for staff and a permanent storage with a capacity of 150 t of potatoes. Also, some storage facilities made from local material have been constructed for tests on how seed potatoes can be cheaply and effectively stored.

Three pickup trucks (2-t capacity) along with three motorcycles (trail) serve as the means of carrying out the program, which includes a considerable amount of travel for the training and extension thrust.

The Belgian government (Administration générale de la coopération au développement) provided 25 million francs (RWF) for the construction of the facilities and provides 5 million RWF for the functioning of the program annually (US $1 = 92 RWF). It is anticipated that the Rwandaise government will match the Belgian contribution. The CIP staff are completely supported by CIP.

Temptations to avoid

Because of its limited facilities and personnel, PNAP has a comparative disadvantage in doing a breeding program. Institutions like CIP and American and European institutions have the personnel and a broad base of genetic material to do an effective breeding program. CIP has as its priorities the development of varieties that are resistant to late blight and bacterial wilt — the two principal problems in Rwanda. Furthermore, it may take one breeder 10 years to develop an improved variety. For example, the USA started a national potato-breeding program in 1929, and researchers there have not yet produced a variety superior to the Russet Burbank, which existed before 1929 and is today the most popular variety.

By receiving each year a large number of CIP tuber families, PNAP staff can screen and find good material, in a short time, with resistance and good yield characteristics; thus, selection is much more practical and appropriate than breeding.

Like a breeding program, a sophisticated virology program is not appropriate at PNAP. In Rwanda, potatoes are grown at elevations higher than 1800 m where viral diseases are secondary in importance. The seed-production program is based on visual screening for viruses. With this approach, farmers can obtain seed of improved varieties that will triple production. A detailed program of using plant tests to screen for viruses would result in the elimination of basic seed potatoes that are much better than what the farmer has at present.

In many developing countries, government seed-certification programs have been attempted, but most have not been successful. The standards set have been too high — thus not realistic. They have led to dishonest practices and, hence, to seed that is not of the quality it is supposed to be. In countries like Rwanda, the stress must be on obtaining seed potatoes that are better than what the farmer has now and supplying him or her with them; demand is so great that, even now, nearly 75%

of the requests for seed potatoes are refused because of a lack of supply. The stress must be on developing a sound reputation as a seed-potato producer and on providing both improved quality and a large quantity to meet the needs of the country.

What is relevant for research depends on the setting of the scientist. In developing countries, it is frequently observed that the desire exists to do research on topics that are not relevant to the constraints to increasing food production. There is peer pressure and pride in doing sophisticated laboratory work. This type of research may be interesting and may produce acceptable publications in international journals, but it does not change the subsistence level of the farmer. For example, the development of nuclear devices by some developing countries at great expense has given them prestige as being technologically advanced; yet, within the same countries, starvation remains a constant reality.

At PNAP, the focus (80–90% of time) must be on the priorities established to improve and increase potato production. However, to ensure that professional interest is maintained and pursued, one may spend 10–20% of the time on research that may not have direct application in the country. Scientists often fail to see the important results of their relevant research and fail to publish them. Thus, the gap between sophisticated research and farmers increases, and the literature does not reflect research that addresses the needs of African farmers.

Few research institutions make direct contact with farmers in developing countries. Meanwhile, the extension worker is blamed for the poor farming practices and the low productivity of the land. Government leaders continually proclaim that the extension worker must show farmers how they can increase production, and the extensionist is blamed for the failure of the farmer to increase production. This failure, however, is often the fault of researchers who:

- Fail to do relevant research that will help the farmers.
- Fail to test the acceptability of their research results in the farmers' fields.
- Fail to inform extension personnel of their research program and train them with the appropriate skills to multiply the successful results obtained.
- And fail to determine the needs of the farmer and resolve them.

The researcher is tempted to avoid the realities of farmers because the research appears routine when in fact it is complex. Agricultural scientists believe this work is for the social scientist, whereas there are real agronomic and pathological questions. Researchers often believe that the extension worker is to help the farmer by adapting research-station results to farmer needs, but the researcher, along with the extension worker, must determine the needs with the farmer and solve the problems as a team. The researcher must be an integral part of technology transfer and not isolated behind the fences of the research station.

Impact

The impact of a development project is often difficult to assess. However, PNAP seems to have made many direct and indirect impacts during its 1st year. For example, 65 t of improved seed potatoes were produced for use in the program and for distribution to seed-potato multipliers. In other words, PNAP has operated the first organized selection program against diseases in Rwanda, using only healthy plants.

Demonstration plots have been established in 13 communities, in farmers' fields. These plots have helped introduce and transfer technology for the selection of healthy plants for seed potatoes, bamboo sticks being used as the markers. The identification of plants infected with virus or bacterial wilt has been demonstrated, as have measures for disease control. Also, the use of fungicides in the control of late blight has been shown on the plots. The best variety currently available, Sangema, was introduced at each site, showing good late-blight resistance and, at the same time, making seed available to the farmer or cooperative. Seven demonstration plots showed that yields could be tripled simply by the use of improved varieties and a fungicide to control late blight. The result has been increased demand for PNAP to produce a large quantity of improved seed potatoes.

Low-cost seed-potato storages have been designed and tested and are now being adopted by farmers and cooperatives who are constructing them with technical assistance from PNAP. Finally, training courses have been provided to 80 extension workers and have been aimed at preparing the trainees to assist the farmers.

These are some of the major direct impacts from demonstrations that will have long-term effects. The research program has, to date, only had an indirect impact but is providing resource information, developing new varieties, and improving agronomic practices that may be transferred in the near future.

Rwandaise farmers have had a direct impact on the PNAP scientists by revealing how challenging it is to produce potatoes with no inputs other than their own resources. The interest and the demand for seed potatoes and training have been overwhelming, suggesting that there is truly a great need for PNAP's existence.

The views and opinions expressed in this paper are my own and do not necessarily represent those of either CIP or ISAR.

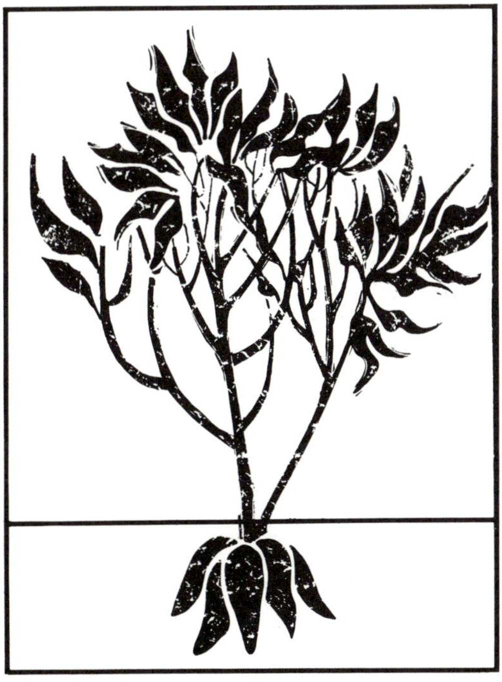

Plant protection

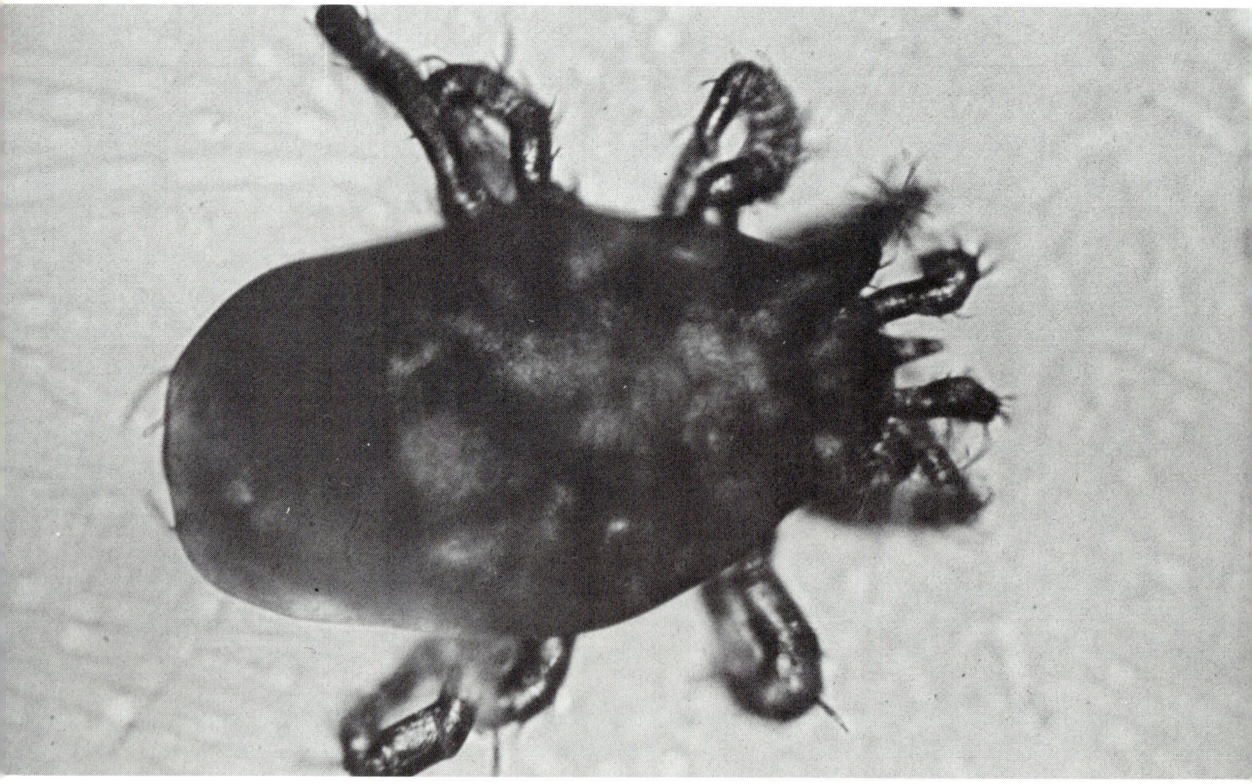

Cassava green mite has spread rapidly since it was first reported in Africa in 1972.

Increasing and stabilizing cassava and sweet-potato productivity by disease resistance and crop hygiene

E.R. Terry and S.K. Hahn[1]

Vegetative propagation encourages the perpetuation and dissemination of the major disease agents affecting cassava and sweet potato. This article describes the major virus and bacterial diseases of both crops with special reference to Africa. Data on crop-yield reductions caused by the diseases are also reviewed. Evidence of host-plant resistance to cassava mosaic and cassava bacterial blight diseases and to the sweet-potato virus disease is presented and reviewed. The possibilities for integrated disease control by the use of the available levels of disease resistance, together with sound crop hygiene, to increase and stabilize the yields of both crops are set forth.

La multiplication végétative est un facteur de perpétuation et de dissémination des principaux agents responsables des maladies du manioc et de la patate douce. L'article décrit les principales viroses et maladies bactériennes qui frappent les deux cultures, surtout en Afrique. On passe en revue les données sur les baisses de rendement dues aux maladies, ainsi que la résistance des plantes hôtes à la mosaïque et la brûlure bactérienne chez le manioc, et à la virose chez la patate douce. On présente les possibilités d'une lutte intégrée contre ces ennemis au moyen des degrés possibles de résistance et d'une bonne hygiène des cultures pour accroître et stabiliser les rendements de ces deux plantes vivrières.

Cassava (*Manihot esculenta*) provides more than 50% of the caloric requirement for 200 million inhabitants in Africa and for 420 million in the world. The total annual cassava production of 4.2×10^7 t in Africa constitutes 40% of the global production (FAO 1977).

Sweet potato (*Ipomoea batatas*) is also an important staple food in many parts of the tropics and subtropics. In addition to its importance as human food, this crop provides animal feed and raw material for industrial purposes. The global production figures show that Africa, producing roughly 4% of the total world output of 1.3×10^8 t, is the second largest producer (FAO 1977).

The incidence and the severity of major economic diseases of cassava and sweet potato limit their productivity both in terms of their high caloric potential and the high protein content of the leaves of both crops, which constitute a significant part of the diet of a large percentage of Africans in Zaire, Sierra Leone, Tanzania, Cameroon, Liberia, and the Congo.

Perpetuation and dissemination of the diseases

Vegetative propagation of cassava and sweet potato provides a highly efficient mechanism for the perpetuation and dissemination of the causal agents of the major economic diseases of the two crops. This is particularly true of the viruses and bacterial pathogens (Terry 1978d).

Of the major virus-like diseases (Table 1) of cassava, the two most important in Africa are cassava mosaic disease (CMD) and cassava brown streak virus (CBSV).

Storey and Nichols (1938) demonstrated that the CMD agent (tentatively believed to be a virus) was carried in cuttings of diseased plants and was transmissible across a graft and by means of an insect vector (*Bemisia* spp.). Storey (1936) first recorded and de-

[1] International Institute of Tropical Agriculture (IITA), P.M.B. 5320, Ibadan, Nigeria.

S03

Table 1. Virus and virus-like diseases of cassava (*Manihot esculenta*).

Disease	Distribution	Causal agent	Dissemination
African mosaic	Africa, India(?)	Paired polyhedral, isometric(?)	Infected cuttings, *Bemisia tabaci*
Common mosaic	Latin America	Flexuous rods	Infected cuttings, mechanical
Brown streak	East Africa	Long flexuous rods(?), paired polyhedral(?), isometric	Infected cuttings, mechanical
Veinal mosaic	Latin America	Polyhedral, isometric	Infected cuttings, mechanical
Latent virus	Latin America, Africa	Rhabdovirus, paired polyhedral	Infected cuttings
Superbrotamento (witches' broom)	Latin America	Mycoplasma	Infected cuttings

Adapted from Lozano (1977).

scribed CBSV in East Africa and demonstrated that the disease was perpetuated through vegetative propagation and that the agent was graft transmissible.

The other viruses and virus-like diseases of cassava are also perpetuated and disseminated by propagation of infected woody stem cuttings (Lozano 1977). Although they are at present restricted in their distribution to parts of South America, the potential for their introduction into the African continent is great.

The major virus diseases (Table 2) of sweet potato are not well understood at present. The relationships between the sweet-potato virus disease (SPVD) complex as described by Schaefers and Terry (1976) and the virus diseases described by Sheffield (1957) and Hollings et al. (1976) remain unclear. Part of the confusion arises from the wide varietal differences in reactions of sweet potato and from complex infections with two or more viruses. Furthermore, some of the viruses involved in these complexes remain inadequately characterized.

Sheffield (1953) reported field observations indicating that the sweet-potato viruses were carried in vine cuttings (primary infection). All the adequately studied sweet-potato viruses have also been shown to be transmitted by grafting, and some are also transmitted (with difficulty) by sap inoculation (PANS

Table 2. Virus and virus-like diseases of sweet potato (*Ipomoea batatas*).

Symptoms	Distribution	Virus particles	Dissemination
Internal cork	USA, South Pacific	Polyhedral, 45–47 nm	Infected vine cuttings, vector
Russet crack	USA, South Pacific	Filamentous pinwheel inclusions	Infected vine cuttings, vector
Witches' broom, "little leaf"	Southeast Asia, South Pacific	Mycoplasma	Infected vine cuttings
Sweet-potato virus disease	Africa, USA	Filamentous rod, 800–950 nm	Infected vine cuttings, vector, mechanical

Adapted from PANS (1978).

1978). Schaefers and Terry (1976) reported two components of the SPVD complex in Nigeria, namely a vein-clearing virus transmitted by aphids in a nonpersistent manner and usually latent in sweet potato and another latent virus transmitted by the whitefly (*Bemisia tabaci*). In combination, the two disease agents produce severe symptoms in sweet potato.

Of the major bacterial diseases of cassava (Table 3), cassava bacterial blight caused by *Xanthomonas manihotis* has been reported from West Africa (Terry 1978c), and both *X. manihotis* and *X. cassavae* have been reported from Kenya (Onyango and Ramos 1978) and Rwanda (Maraite and Perreaux 1978).

Xanthomonas manihotis is a vascular pathogen that can be disseminated through vegetative planting material (Lozano and Sequeira 1974). Although *X. cassavae* was originally reported as inducing only leaf spots (Wiehe and Dowson 1953), it is also possibly disseminated by vegetative propagation of infected woody stems. *Xanthomonas manihotis* also could be disseminated by movement of soil during cultural operations or by the use of contaminated tools, especially in view of the extensive amount of cutting that is required during harvesting and preparation of planting material (Lozano 1975). Studies on the role of insects in the dissemination of this pathogen revealed that spread attributable to insects only occurred over short distances (Lozano and Sequeira 1974).

The two other cassava bacterial pathogens, *Erwinia carotovora* var. *carotovora* (bacterial stem rot) and *Agrobacterium* sp. (bacterial stem gall), have been reported only in the Americas but have a high-risk potential for introduction into Africa as well.

Two bacterial pathogens have been reported for sweet potatoes: *Pseudomonas batata*, which causes a wilt in China (Faan et al. 1959), and *Erwinia* spp. causing wilt and plant death in the United States (Martin and Dukes 1975). Both of these bacteria are disseminated through diseased vines (Nielsen and Terry 1977).

Yield losses

Cassava mosaic disease

The symptoms of CMD are characteristic of a mosaic disease, primarily chlorosis of discrete areas of the leaf lamina. Leaf-area reduction is reported to be about 24% (Beck and Chant 1958), and severely affected leaves are twisted and distorted. Cassava mosaic disease induces increases in the respiration rate and in the peroxidase activity in cassava leaves and reduces photosynthetic activity of mature infected leaves by 23%. Chloroplasts in the mesophyll cells of infected leaves are irregularly shaped and contain numerous swollen starch grains (Chant et al. 1971).

Jennings (1972) estimated that the yield losses caused by cassava mosaic in some locales in East Africa ranged from 65% to 95%. Bock and Guthrie (1978) reported that, in Kenya, the mean loss from CMD in a moderately resistant hybrid was 70%, and, in a

Table 3. Bacterial diseases of cassava (*Manihot esculenta*).

Disease	Global distribution	Causal agent	Dissemination
Cassava bacterial blight	Latin America, Africa, Asia	*Xanthomonas manihotis*	Infected cuttings, rain splashing
Cassava bacterial leaf spot	Africa	*X. cassavae*	Infected cuttings(?), rain splashing
Cassava bacterial stem rot	Latin America	*Erwinia carotovora*	Insects
Cassava bacterial stem gall	Latin America	*Agrobacterium* spp.	Infected cuttings, infected soil

Adapted from Lozano (1977).

susceptible cultivar, it was 86%. We (1980) found that, in Nigeria, the mean fresh-root weight reduction caused by CMD ranged from 32% to 69% and that the starch-content reduction was 2.5–7%.

Cassava brown streak virus

Symptoms of CBSV occur in leaf, stem, fruit, and root tissue. The main leaf symptom is chlorosis, whereas stem symptoms consist of brown streaks that elongate and coalesce with adjacent streaks to form blotchy patches. Necrotic lesions can also be observed in leaf scars after leaves have been shed. Root infection results in the formation of necrotic lesions in the tuberous starch-storage tissues (Nichols 1950).

Because CBSV and CMD invariably occur together in the coastal regions of East Africa, it has not been possible to determine the yield loss owing to CBSV alone; however, the extensive root necrosis renders tuberous roots unfit for human consumption (Jennings 1972).

Cassava bacterial blight

The characteristic symptoms of CBB include angular water-soaked leaf spots, which are initially small but later enlarge, coalesce, and eventually turn brown; various degrees of leaf wilt; the exudation of a yellow-orange gum on the leaves, petioles, and young shoots; severe defoliation; and finally stem dieback resulting from vascular necrosis and death of growing tips.

Cassava bacterial blight is considered the most devastating of several bacterial diseases of cassava because it can result in total loss of both yield and planting materials of a susceptible variety (Lozano and Sequeira 1974). Under favourable conditions for disease development in areas where the leaves are consumed, the heavy loss of foliage constitutes a significant economic loss (Terry 1978c). Tuberous root-yield losses have been reported as 7–75% in Colombia (CIAT 1975); 58% in Ibadan, Nigeria (IITA 1977); and 90% in Umudike, Nigeria (O.B. Arene, unpublished). Reduction of tuberous-root starch content because of CBB has also been reported (Obigbesan and Matuluko 1977).

Sweet-potato virus disease

Symptoms of the sweet-potato virus disease consist of various combinations of leaf strapping, vein clearing, puckering, and stunting (Schaefers and Terry 1976). Such visible symptoms in virus-infected plant tissues are usually the result of disturbances in host metabolism, increased respiration and peroxidase activity, and decreased photosynthetic activity, which, together, lead to considerable reductions in yield.

Sheffield (1953) reported that the disease was often severe in the west of Uganda, in Rwanda-Urundi. In Kenya and Tanzania, although the disease was widespread, it apparently never reached serious proportions except in a few places such as Kisumu. Mukiibi (1977) reported that the disease caused a reduction of 57% in yield, in terms of both weight and number of tubers of the sweet-potato variety Kyebandula. In Nigeria (IITA 1974), it was reported that, in trials using slips from plants with and without symptoms, yield of fresh tubers was reduced by 78–80% in plants with SPVD symptoms but that the reduced chlorophyll level of diseased leaves did not affect the yield.

Disease control by host-plant resistance

Cassava diseases

The occurrence of linked resistances to CMD and CBB was first reported by Hahn (IITA 1973). The resistance to CMD that was inherited from the *M. glaziovii* derivative 58308 was largely conferred by recessive, additive genes and was associated with resistance to CBB. Resistance to both CMD and CBB has been successfully incorporated into high-yielding cultivars of acceptable quality, and the improved varieties have been evaluated over the last 6 years. Many promising clones from IITA-resistant material have been selected in various countries in tropical Africa and India.

With regard to CMD, it is difficult to postulate the mechanism of resistance because its causal agent remains unidentified. However, resistant cultivars have shown significantly lower CMD severity ratings and only small percentages of infected plants for long periods under heavy disease and *B. tabaci* pressure. This suggests that the rate of disease spread is limited in plants of the resistant cultivar (Hahn et al. 1980).

The CMD-resistant breeding materials developed at IITA were tested during 1973–78 in West, Central, and East Africa and proved effective. The absence of regional differences in terms of effectiveness of this resistance and its polygenic nature suggest that it is durable over time and distance (Hahn et al. 1980). Whether the resistance is race-nonspecific is not yet clear, but information on pathogenic variation should provide an answer.

With regard to CBB, the high level of resistance of IITA-improved varieties to this disease was confirmed by stem-puncture, leaf-clipping, and leaf-spray inoculation tests (Perreaux et al. 1978) and is well correlated with the results of natural infection (IITA 1976). Highly tolerant varieties only exhibit angular leaf spotting under natural infection and rarely show the degree of defoliation and stem dieback associated with susceptible varieties. Resistant varieties may have a relatively high number of leaf spots per leaf, but these are normally very small and do not coalesce to form large spots.

Sweet-potato virus disease

Hahn et al. (1981) reported that, in field trials, in four locations in Nigeria, four sweet-potato cultivars showed high levels of field resistance to SPVD. A high correlation between field SPVD severity scores and scores recorded after a positive challenge by the root-graft transmission method confirmed the resistance level of the IITA-improved varieties.

Individual plants of the resistant varieties were challenged with virus-carrying core tissues, and, although the test varieties expressed symptoms, there was evidence that these varieties had the capacity to grow vigorously and recover from infection. Only small numbers of plants of resistant cultivars were positively challenged by this method, and there were significant differences in time interval for onset of symptoms (Terry 1979). These observations indicate that the nature of resistance is associated with resistance to infection.

Disease control by crop hygiene

Results of field observations and experiments in Kenya suggest that CMD can be controlled by the use of mosaic-free planting material. The low rate of spread of mosaic into mosaic-free plots, and also within plots, indicates that whiteflies are comparatively inefficient vectors and that human beings are the principal vector because of their indiscriminate use of infected cuttings as propagation material (Bock and Guthrie 1978). Lozano and Terry (1976) suggested that the incidence of pathogens and the damage they cause to cassava propagation material could be reduced by a careful selection of all planting material. The use of cuttings from plantations with CBB infection must especially be avoided, and any stem sections with signs of disease should be eliminated.

Sheffield (1953) reported similar findings for sweet potato — that the increase of SPVD by the vector was normally slight in the field — and suggested that the more important damage arises from propagation by infected vines. The merits of utilizing pathogen-free planting material especially for vegetatively propagated crops, to minimize crop losses, are well recognized (Hollings 1965). The strategy is to dilute the amount of virus or bacterial infection by the supply of large quantities of healthy planting material to facilitate replacement of reinfected stock.

Integrated disease control

The success of any control practice based on utilization of pathogen-free planting material depends largely on the incidence and magnitude of subsequent field reinfection by the major pathogens of the crop. Subsequent field reinfection is influenced in turn by the varietal susceptibility of the planting material, the virulence of the relevant pathogens, and the existence of favourable environmental conditions for disease development.

We (1980) reported earlier that the improved CMD-resistant cassava variety TMS 30395 produced significantly higher root yields under field conditions than its susceptible parent Isunikakiyan irrespective of whether it was established from mosaic-free or mosaic-infected planting material. Under conditions in Ibadan, Nigeria, the CMD-resistant TMS 30395 yielded an average of 2.96 kg fresh roots/plant when established from CMD-free cuttings compared with 2.02 kg/plant from CMD-infected cuttings. The CMD-susceptible variety Isunikakiyan yielded 1.43 kg/plant when established from CMD-free cuttings, compared with 0.44 kg/

plant from CMD-infected cuttings. Furthermore, the rate of increase in CMD incidence on TMS 30395 established from CMD-free planting material was low, reaching a peak of 17.4%, 5 months after planting, compared with 99% for Isunikakiyan. It appears, therefore, that an important component of resistance to CMD is resistance to reinfection by the *B. tabaci* vector.

This type of resistance, which results in low field incidence of CMD in resistant varieties like TMS 30395, can be exploited for the possibilities it offers for maintaining a CMD-free crop yielding an average of 29 t/ha, the farmer roguing out the small number of plants that become infected. These plants can then be routinely replaced by CMD-free plants of the same cultivar from a disease-free stock.

Hahn et al. (1981) reported that the IITA-improved sweet-potato varieties resistant to SPVD can also yield an average of 20–30 t of roots/ha in 4 months without fertilizers. Furthermore, in field tests, one of the resistant varieties, TIS 2498, had a low SPVD severity rating and only a small increase in incidence.

The aphid-transmitted filamentous particles of the SPVD in Nigeria can, with difficulty, be mechanically transmitted from sweet potato to *Nicotiana benthamiana*, inducing leaf crinkling and puckering, but not to sweet potato (IITA 1978). In East Africa, a virus obtained from sweet potatoes in Kenya, Uganda, and Tanzania was transmitted by inoculation of sap and by whiteflies (*B. tabaci*) to sweet potato and other plant families (Hollings et al. 1976). In Nigeria, however, although large populations of whiteflies and aphids have been reared in captivity on sweet potatoes, field infestation levels are rather low, and it appears that field-transmission efficiencies of SPVD by these vectors are also quite low (Hahn et al. 1981). In Ibadan, Nigeria, it normally takes 1–3 years for seedlings to develop severe SPVD under field conditions. Hahn (1979) reported that most of the symptomless plants used as disease-free controls in trials for yield comparison between SPVD-infected and disease-free plants remained symptomless until harvested 4 months after planting. The evidence, therefore, indicates that farmers, by propagating vine cuttings from infected sweet-potato plants, are the principal vector of SPVD.

The low rate of disease spread in disease-free plants of resistant varieties offers the possibility for farmers to maintain an SPVD-free crop, yielding an average of 20–30 t/ha, and to rogue out the small number of plants that become infected.

Effects of soil fertility on cassava bacterial blight in Rwanda

I. Butare and F. Banyangabose[1]

This paper discusses the ecological factors that influence the seriousness of bacterial blight caused by *Xanthomonas cassavae* in Rwanda. Field observations have indicated that the disease is most serious where the soil is poor, gravelly, and sandy; therefore, experiments were conducted to determine the effect of soil fertility on bacterial blight.

Discussion des facteurs écologiques qui influent sur la gravité de la brûlure bactérienne due à *Xanthomonas cassavae* au Rwanda. Les observations sur le terrain ont révélé que la maladie prolifère sur les sols pauvres, pierreux et sablonneux; on a donc fait des expériences pour déterminer l'impact de la fertilité du sol sur la brûlure bactérienne.

In Rwanda in 1977 and 1978, cassava was severely damaged by bacterial dieback caused by *Xanthomonas cassavae*. Studies conducted at that time in the main cassava-producing regions showed that bacterial blight was most prevalent in areas where the soil was poor, gravelly, and sandy. Although the disease has since receded in all areas of Rwanda, an in-depth study of the ecological factors influencing the seriousness of the disease was initiated. This paper summarizes the experiments conducted to determine the effects of soil fertility on bacterial blight.

Soil samples were taken from the eight main centres of bacterial blight and analyzed in the soil chemistry laboratory at Institut des sciences agronomiques du Rwanda (ISAR). Both particle-size and chemical analyses were carried out (Tables 1 and 2). Mineral deficiencies were determined by means of the pot-culture method with procedures normally used in ISAR's chemistry laboratory.

Textural analysis has indicated that the areas in which bacterial blight is most likely to occur are those with sandy-clay soils (Gitovu, Kavumu, Cyayi, Gikonko, Nzuki) or clayey-sand soils (Nyamiyaga, Musamo, Ruyenzi). Most of the soils (Gitovu, Kavumu, Nyamiyaga, Musamo, Nzuki, Ruyenzi) are deficient in calcium and magnesium and have low exchangeable-base levels. All of the soils (except those in Cyayi) have low potassium levels. Assimilable phosphorus is also very low.

Studies of mineral deficiencies based on pot

[1] Institut des sciences agronomiques du Rwanda (ISAR), B.P. 138, Butare, Rwanda.

Table 1. Particle size analysis — a percentage based on INEAC's soil-texture triangle.[a]

	Particle size (μm)			
	0–20	20–200	200–2000	Soil type
Gitovu	35.6	26.8	37.6	Sandy clay
Kavumu	41.8	9.0	49.2	Sandy clay
Nyamiyaga	23.4	22.4	54.2	Clayey sand
Musamo	22.0	19.0	59.0	Clayey sand
Cyayi	43.2	8.6	48.2	Sandy clay
Gikonko	38.8	15.6	45.6	Sandy clay
Nzuki	35.8	13.0	51.2	Sandy clay
Ruyenzi	29.6	23.6	46.8	Clayey sand

[a]0–20 μm is fine elements, 20–200 μm is fine sand, and 200–2000 μm is coarse sand.

Table 2. Chemical analyses of soils from different sites.

	Organic matter			pH		Absorbing complex (meq/100 g)				Assimilable
	C %	N %	C/N	H_2O	KCl	Ca	Mg	K	Na	P (ppm)
Gitovu	1.40	0.15	9.33	5.20	4.00	1.10	0.51	0.18	0.05	trace
Kavumu	3.09	0.29	10.65	5.50	4.70	0.24	1.88	0.38	0.10	2
Nyamiyaga	1.97	0.17	11.59	5.50	4.45	0.57	0.11	0.15	–	2
Musamo	1.41	0.11	11.82	4.60	4.20	1.13	0.33	0.20	0.03	5
Cyayi	3.59	0.33	10.88	5.20	4.95	7.28	2.28	0.65	0.03	3
Gikonko	4.23	0.43	9.84	4.80	4.70	9.75	2.63	0.24	0.04	3
Nzuki	1.69	0.17	9.94	4.10	3.90	0.47	0.17	0.25	0.03	2
Ruyenzi	1.75	0.14	12.50	5.00	3.90	0.92	0.23	0.26	0.07	2

culture usually assume that dry-matter production of less than 40% indicates high levels of deficiency; production levels of less than 20% indicate extreme deficiency. In all cases, the production of dry matter in the control pots was low; therefore, it can be concluded that the general fertility level of these soils is low. All the soils studied were highly potassium and phosphorus deficient.

Effect of soil fertility on disease development

Different concentrations of a nutrient solution were added to the soil at Kavumu. The method employed is based on the premise that a soil that has received 100 ml of the solution of major elements and 10 ml of the solution of microelements is in equilibrium in terms of fertility. The composition of mineral elements was (in mg/kg of dry soil): CaO 210, Mg 100, K_2O 590, S 120, P_2O_5 495, N 600, Cu 0.36, Mo 0.21, Mn 2.14, CO 0.07, Fe 0.07, Zn 0.64, B 1.43, and Na 1.43.

Different fertilizer treatments were used (Table 3). The soil was soaked, and the leachate was periodically recovered and poured back into the pot. The cassava variety used was Eala Amer 07, 11-week-old cuttings being inoculated by a needle contaminated with a 5-day-old bacterial culture. In each group, a plant that had not been inoculated was used as a control. The experiment examined the number of diseased plants, the length of the necrotic zone on the stem, and the number of plants that suffered from dieback.

Table 3. Treatments used to determine effect of soil fertility (15 pots used in each treatment).

	Quantities added (ml)	
Treatment	Macroelements	Microelements
FC[a]	180	20
½ FC	90	10
¼ FC	45	5
Control	0	0

[a]FC = full concentration.

Results

The number of plants showing symptoms of necrosis on the stem was observed 7 and 11 days after inoculation (Table 4). Neither the time nor the soil treatment made a significant difference in the numbers of plants infected. In other words, there was no relationship between the incubation of the disease and the increase in soil fertility. However, after 3 weeks, an analysis of variance, after the data had been converted into a log form (x + 1), revealed that plants that received the full concentration (FC) differed significantly from control plants and from plants that received ¼ FC (Table 5). In other words, the development of the disease in plants that received FC was slower than in plants that received the other treatments. After 4 weeks, this difference became very significant. However, there was no significant difference between plants receiving FC and ½ FC or among those receiving ½ FC, ¼ FC, and no fertilizer (control). It can, therefore, be concluded that an increase

Table 4. Number of plants showing symptoms of necrosis.

	Plants attacked		Plants not attacked	
Treatment	7 days	11 days	7 days	11 days
FC[a]	8	9	6	5
½ FC	6	8	8	6
¼ FC	5	10	9	4
Control	8	10	6	4

[a]FC = full concentration.

Table 5. Length (cm) of the necrotic zone on the stem, 3 and 4 weeks after incubation.

FC		½ FC		¼ FC		Control	
3 weeks	4 weeks	3 weeks	4 weeks	3 weeks	4 weeks	3 weeks	4 weeks
1.6	2	0.9	1.1	2.05	4.5	2.4	8
2.8	4.2	1.1	1.3	2.6	5	3.5	10.7
2.3	2.6	3.3	5.8	2.5	3.6	2.8	6.7
2.1	3.5	1.4	2.1	4.4	6.8	1.8	3.4
1.8	2.6	1.3	1.8	2.2	5.8	0	0
1	1.6	1.9	2.9	2.8	3.5	0	0
1.4	1.4	1.2	2.3	3.1	9.2	3.3	6.8
2.0	2.3	2.0	3.8	2.6	5.0	7.1	10.5
2.4	3.8	2.4	3.8	3.6	7.4	3.7	7.5
0	0	1.0	1.9	0	0	3.5	7.2
0	0	3.1	6.6	2.5	4.9	5.2	7.7
0	0	3.1	6.1	2.0	6.9	0	0
0	0	1.5	2.4	2.4	3.9	2.8	5.7
0	0	0	0	1.9	2.8	2.8	5.7

in the general level of soil fertility hinders the development of the disease.

After 5 or 6 weeks, the results were not significantly different at the 0.05 level. The number of plants attacked by the disease was, therefore, independent of the treatments. However, an increase was observed in the incidence of dieback in both the plants that received ¼ FC and the control plants. None of the plants receiving the full concentration suffered from dieback.

Conclusions

It can be concluded from these experiments that:

- The areas in which bacterial blight is most prevalent are those with sandy-clay soils or clayey-sand soils.
- The areas in which the disease is most common all have poor soil (low level of fertility). In addition, the soils are phosphorus and potassium deficient.
- An increase in the general level of fertility through the addition of nutrient elements will not prevent infection but will hinder the development of the disease.
- The fact that cassava is grown mainly in areas with poor soil contributes to the high incidence of bacterial blight.
- Potassium and phosphorus probably play an important role in the resistance of cassava to bacterial blight. A factorial analysis for phosphorus and potassium would determine this role more precisely.

We would like to thank our colleagues in the Pedology Division who greatly helped us in conducting the experiments and analyzing the results.

Distribution and importance of *Xanthomonas manihotis* and *X. cassavae* in East Africa

D.M. Onyango and D.M. Mukunya[1]

The diseases caused by *Xanthomonas manihotis* and *X. cassavae* are currently referred to as cassava bacterial blight. Although the symptoms of the diseases do overlap, some distinctive features can be observed, and the bacteria can be identified on the basis of the symptoms of the diseases they cause. Both species have been reported in East Africa, probably introduced from West Africa via Central Africa. However, surveys (1976–77 and 1978–79) have indicated that *X. manihotis* is probably more important than *X. cassavae* in the region.

On appelle communément brûlure bactérienne les maladies du manioc causées par *Xanthomonas manihotis* et *X. cassavae*. Bien que les symptômes de ces maladies se recouvrent partiellement, elles ont chacune des caractères spécifiques qui permettent d'identifier la bactérie responsable. On a signalé la présence de ces deux bactéries en Afrique orientale, venant probablement de l'ouest en passant par le centre du continent. Par ailleurs, les investigations conduites dans la région en 1976–1977 et en 1978–1979 ont révélé un degré de virulence plus élevé dans le cas de *X. manihotis*.

Xanthomonas manihotis and *X. cassavae* are among the five bacteria that are known to cause disease in cassava. The diseases caused by these two bacteria are currently being referred to as cassava bacterial blight (CBB). Although the symptoms for the two diseases do overlap, some distinctive features can be observed. Because both pathogens do occur in Kenya, careful observations have been made so that the two bacteria can be identified on the basis of the symptoms of the diseases they cause. Both bacteria have been noted to cause angular water-soaked spots on the leaves. Careful observation has shown that, for *X. cassavae*, spotting is concentrated along the midrib and lateral veins, with a few scattered spots on the lobe (E.R. Terry, personal communication). The spots for *X. manihotis* have not been noted to concentrate along the veins.

Both bacteria cause defoliation, the severity of which depends on varietal susceptibility. Generally, in the case of *X. manihotis*, there is more defoliation when infection originates from the leaves than in the case of systemic infection. Both bacteria cause exudation on leaf lobes (lower side), petioles, and young shoots. Exudates of *X. manihotis* are initially white and fluidy but turn yellow-orange on drying. Their size also depends on varietal susceptibility. Exudates of *X. cassavae* are generally very tiny and are golden-yellow on drying.

Only *X. manihotis* causes typical vascular wilt of leaves and shoots. The pathogen also causes tip dieback of shoots. These two features are characteristic of *X. manihotis*. *Xanthomonas cassavae* causes neither vascular wilt nor tip dieback. Death, however, results from severe necrosis of tissues at the site of infection. Therefore, all parts of the shoot distal to the point of infection will dry and die, following the disconnection at the necrotic site from the rest of the plant. We suggest that two names be adopted for the two diseases to eliminate the confusion that prevails at the moment. Cassava bacterial blight should be reserved for the disease caused by *X. manihotis*, whereas the disease caused by *X. cassavae* should be referred to as cassava bacterial necrosis. One name should not be used for two diseases caused by two different

[1] National Agricultural Laboratories (NAL), P.O. Box 14733, Nairobi, Kenya, and University of Nairobi, P.O. Box 30197, Nairobi, Kenya, respectively.

pathogens. The use of such a common name becomes particularly inadequate when reference is being made to both pathogens.

In East Africa, *X. manihotis* has been reported in Kenya, Uganda, and Tanzania; *X. cassavae* has been observed in a number of countries in East and Central Africa: Uganda, Malawi, Kenya, and Rwanda.

Detailed surveys were carried out in 1976–77 and 1978–79 in all cassava-growing areas of Kenya. The methods and isolation procedure adopted were those of Onyango and Ramos (1978). Confirmation of isolates as *X. manihotis* or *X. cassavae* was done by pathogenicity tests in varieties F279, Dodo, and 5543/156. All inoculations were done in greenhouses with plants 2–3 months old. The surveys were extended to the borders of Tanzania and Uganda.

Results

In Kenya, *X. manihotis* was observed in the entire western side, from Tanzania's border to Uganda's border. The pathogen was more prevalent in the region north of Lake Victoria than in the south. It was not observed in the regions east of the Rift Valley. The surveys indicated that *X. cassavae* was widely distributed in western Kenya, extending between the boundaries with Uganda and Tanzania. The prevalence of the disease was high (up to 60%) around the lake region. It was noted that the prevalence and the severity of *X. cassavae* were always lower than those of *X. manihotis*. Both bacteria were found from Tanzania's border to Uganda's border and were confined to western Kenya.

Discussion

The presence of *X. manihotis* in nearly all cassava-growing countries of the world probably can be explained by a common origin for cassava. Cassava, thought to have been domesticated more than 5000 years ago, originated from the Americas, with a major centre of diversity in Brazil and a secondary one in Mexico. We believe that cassava was distributed from Brazil across the Atlantic Ocean to the African continent and beyond to Asia by the Portuguese.

Xanthomonas manihotis is thought to have existed in Brazil as early as 1911. The wide distribution of the pathogen in individual countries in East Africa suggests its long establishment in the region. Its spread within and across the states was effected by transfers and exchange of infected cassava materials. Confinement of this pathogen in western Kenya and not at the coast has confirmed the important role of transfer of diseased material in long-distance dissemination.

The losses reported in Tanzania, Uganda, and Kenya have shown the importance of the disease in the region. Continuous use of common varieties in provinces and in overlapping zones at the borders has resulted in epiphytotics. In Kenya, susceptible varieties such as Dodo and Tamisi (local) and F279 and F100 (introduced) were completely wiped out during the 1976–77 heavy rains. Greenhouse pathogenicity experiments conducted in 1979 confirmed the field observations regarding the susceptibility of these varieties.

We believe that *X. cassavae* occurs widely in Tanzania, especially around Lake Victoria and the Mara region. Our 1979 survey indicated that the pathogen extended into Tanzania across the border near Sirari. Its occurrence in wide and isolated areas in Kenya suggests how established it must be. The higher incidence around Lake Victoria and toward the south (Tanzania's border) and its occurrence in Malawi and in Uganda suggest a distribution pattern.

Its frequent association with isolated *Manihot glaziovii* trees confirms the pathogen's ability to disseminate itself. Up to the mid-1960s, there was free and frequent movement of materials among the East African states. Under such circumstances, the two bacteria were easily disseminated. The centrally situated Lake Victoria connected the three states by way of trade and communications. Thus, it would be a surprise if *X. cassavae* did not occur in all three states around the lake.

Xanthomonas cassavae is unlikely to be an important pathogen in East Africa. Survey data showed that, even in the areas of highest incidence, the pathogen caused little crop loss. The pathogen was ineffective against mature plants, although in Rwanda serious disease caused by *X. cassavae* has been noted. This observation indicates that the pathogen should not be dismissed as unimportant. Under certain conditions, yet unknown, it can express itself very strongly. Nevertheless, although *X. cassavae* can cause leaf defoliation and death of young shoots by girdling, damage from it was not found to be serious in the survey. The pathogen can be disseminated as easily as *X. manihotis* but is nonsystemic and produces very little exudate, if any, on cassava stems.

The spots, especially during the rainy season, are the major source of inoculum. The slow establishment of *X. cassavae* in the host and the lower maximum temperature it can tolerate are further disadvantages to the bacterium.

The presence of both bacteria in Rwanda and Burundi cannot be ruled out. We feel that both bacteria have probably always existed in most cassava-growing areas and that the two pathogens are much older than is reported, considering their wide distribution. Finally, we are of the opinion that these bacteria, especially *X. manihotis*, were introduced into East Africa from West Africa via Central Africa. Evidence for this hypothesis is that:

- More damage by disease has been reported in West and Central Africa than in East Africa;
- Cassava has been and remains more important in West than in East Africa;
- The west coast of Africa had more direct contact through the Portuguese with Brazil than did East Africa;
- In East Africa, the disease incidence decreases from the west toward the east (in Kenya, there is no disease at the coast);
- The disease has been reported in a continuous zone from the west coast of Africa through the Central African countries to the west of Kenya; and
- Rapid dissemination of *X. manihotis* to distant places through infected materials is well documented.

Cassava mosaic disease

E.J. Guthrie[1]

A geminivirus has been isolated from cassava infected with cassava mosaic disease (CMD) on the Kenyan coast; it is related but distinct from an earlier isolate from western Kenya and requires different procedures for isolation. This newly isolated virus is probably a causal agent of CMD in eastern Kenya, although it may not be the sole cause.

Un virus bivalent a été isolé sur des plants de manioc atteints de mosaïque, sur la côte du Kenya; il s'apparente, bien que distinct, à un virus déjà isolé dans l'ouest du pays ; cependant, on ne réussit à l'isoler qu'au moyen de techniques différentes. Ce nouveau virus pourrait être l'un des agents responsables de la maladie dans l'est du Kenya.

In Kenya, a geminivirus was isolated from cassava into *Nicotiana clevelandii* (Bock et al. 1978). As the virus was apparently confined to western Kenya and could not be isolated from cassava infected with mosaic disease (CMD) at the coast, it was referred to as cassava latent virus (CLV).

More recently, a geminivirus has been obtained from CMD-infected cassava at the coast, *N. benthamiana* being the isolation host. The method of purification necessary differed from that used in isolation of the virus from west Kenya, and serological tests show the two isolates to be related but distinct. A CMD isolate from Nigeria was apparently identical to the west Kenya type.

It seems possible that the newly isolated geminivirus is the causal agent of CMD, although the existence of another component cannot be ruled out.

Cassava originated in the New World, where CMD does not occur — a fact that suggests that there is an alternative host or hosts for the virus in Africa. The virus has recently been isolated from *Jatropha multifida* (Euphorbiaceae) and *Hewittia sublobata* (Cucurbitaceae); *H. sublobata* is widespread in the Old World tropics and may well prove to be significant in the ecology of the disease.

The available information on the distribution of the two strains of CMD is in line with theories that cassava arrived in East Africa by two routes: to the west from West Africa and to the coast from the offshore islands. Further sampling, particularly in Tanzania, would be instructive.

[1] Kenya Agricultural Research Institute, P.O. Box 30148, Nairobi, Kenya.

Pest control for cassava and sweet potato

K. Leuschner[1]

The prevailing socioeconomic conditions of the smallholder farmer in Africa militate against the widespread use of insecticides not only because of the costs of the chemicals but because of the long-term side-effects from their misuse. Thus, other means of pest control must be sought. Agricultural researchers must concentrate on building-in host-plant resistance, identifying and introducing natural enemies of the offending pests, and developing agronomic practices that discourage the buildup and spread of the pests. Progress along these lines for two important staple root crops — cassava and sweet potatoes — in Africa has been considerable over the last 10 years; the focus and accomplishments of work to date are reviewed in this paper.

La situation socio-économique actuelle des petits exploitants africains empêche le recours généralisé aux insecticides, tant à cause du coût des produits chimiques que des effets secondaires à long terme de leur mauvaise utilisation. Il est donc impérieux de trouver d'autres formes de lutte. Les chercheurs agricoles doivent se concentrer sur le renforcement de la résistance des plantes hôtes, l'identification et l'introduction des ennemis naturels des insectes et le développement de pratiques agronomiques qui puissent enrayer l'accroissement et la propagation des insectes. On a fait des progrès considérables dans ces domaines pour deux importantes racines alimentaires de base — le manioc et la patate douce — en Afrique au cours des dix dernières années ; l'article passe en revue les objectifs et les résultats des travaux réalisés jusqu'à présent.

Tropical root crops contribute substantially to the carbohydrate diet of people in West, Central, and East Africa. Cassava is the most important crop in Central Africa and parts of West Africa. Sweet potatoes are very important in East Africa and, to a lesser extent, in the rest of Africa. According to the Food and Agriculture Organization of the United Nations (FAO 1976), Africa is the biggest cassava producer in the world.

In Africa, cassava and sweet potatoes are mainly grown together with maize, beans, and vegetables. The crop varieties grown under these conditions are generally poor yielding but fairly well adapted to the ecological conditions and pest fauna. With the increasing demand for food, improved agricultural practices like monocultures and high-yielding varieties are slowly being adopted. This trend certainly will change, and is already changing, the pest status in these crops. One has to anticipate the resulting problems by developing pest-management programs for these crops.

Knowledge of insects and mites attacking these crops is still limited. Although the main pests have been identified, the basic biological and, in many cases, yield-loss data are lacking.

This paper presents some of the available knowledge, together with control approaches that might work under African conditions.

Cassava pests

Until recently, cassava in Africa was relatively free of major pests. The reason probably was that most of the pest species attacking cassava in South America were not introduced with the plant, and relatively few African insect and mite species have adapted to cassava, probably because of its cyanide content.

Until 1970, only the whitefly (*Bemisia tabaci*), the vector of CMD; the red spidermite (*Tetranychus cinabarinus*); the white scale

[1] International Institute of Tropical Agriculture (IITA), P.M.B. 5320, Ibadan, Nigeria.

(*Aonidomytilus alleus*); and a grasshopper (*Zonocerus variegatus*) were damaging cassava. *Zonocerus* and *Tetranychus* defoliate cassava during the dry season, starting from the lower old leaves and progressing slowly to the top of the plant. With heavy attack and repeated defoliation, damage from *Zonocerus* may reach 60% (COPR 1977).

The whitefly itself does not cause much damage to cassava, but whitefly density in the field is highly correlated with the incidence of cassava mosaic disease (Fig. 1). White scale attacks the cassava stems and may prevent sprouting if infested stakes are used as planting material. At present, all these pests are controlled by natural enemies.

However, this is not the case for two pest species introduced into Africa about 1970 by vegetative planting material from South America: the cassava green mite (*Mononychellus tanajoa*) and a mealybug (*Phenacoccus manihoti*).

The green mite was first observed in Uganda (Nyiira 1972) and, since then, has spread to Kenya, Tanzania, Burundi, Rwanda, Zaire, Nigeria, Benin, and Togo. It is a dry-season pest that attacks cassava shoots. The symptoms are yellowing of young leaves and inhibition of new leaf development. One generation takes about 10–12 days. Adults are carried by wind over long distances. Yield losses up to 40% were reported by Nyiira in 1975.

Phenacoccus manihoti was first observed around Kinshasa (Zaire) (Leuschner 1976) and, since, has spread to Angola, the Congo, Gabon, Senegal, Gambia, Nigeria, Benin, and Togo. Like the green mite, the mealybug attacks the shoot, prevents new leaf growth, and defoliates the plant in the late dry season. The insect is parthenogenetic, and one life cycle is completed in about 20 days. The pest disperses by wind (crawler stage) and on cuttings. It results in yield losses from 20% to 50% and may kill young plants. Both pests decrease to subeconomic levels during the wet season.

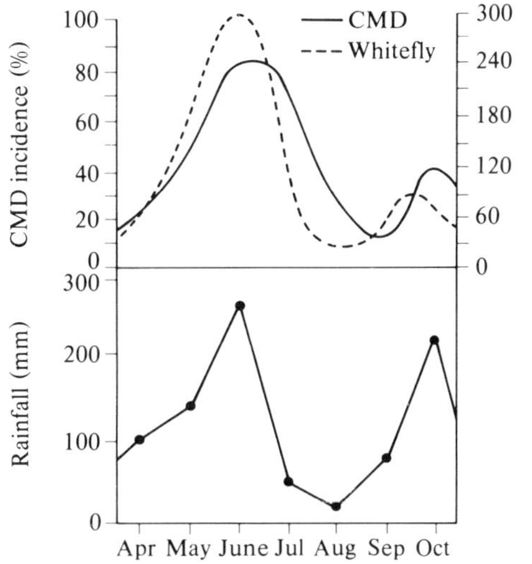

Fig. 1. Whitefly population development compared with CMD incidence and rainfall pattern over 6 months.

Sweet-potato insects

The main harmful insects on sweet potatoes in Africa are the weevils *Cylas formicarius* and *C. puncticollis*. The former is mainly distributed over East Africa, whereas the latter is a West and Central Africa species. The weevils are similar in their behaviour: they attack the leaves, the stem, and the sweet-potato tuber. Females lay about 200 eggs, and the life cycle is 20 days. Economic damage is attributable to tuber attack and, to a lesser extent, to stem attack (Cockerham et al. 1954). The weevil population increases during the dry season because of higher temperatures and soil cracks that expose the tubers. The larvae not only tunnel into the tuber but also deposit wastes that give the tuber a disagreeable taste.

In addition, secondary fungus and bacterial infections arise and destroy the tuber. Loss of tubers has been reported as high as 90% in India (Fröhlich and Rodewald 1970) with high weevil infestations, but, normally, losses range between 10% and 50%. The damage continues during storage.

Minor pests are the sweet-potato hornworm (*Agrius convolvuli*) and a tortoise beetle (*Conchyloctaenia punctata*). Both are leaf feeders, their contribution to yield loss being unknown at present (PANS 1978).

Because of the low educational standard of African farmers, lack of money, irregular supplies of insecticides, and the relatively undisturbed agroecosystems in Africa, measures to control pests should be well planned and should only include the use of insecticides when no other control is feasible. This means that the three fundamental tactics should be:

- To build-in host-plant resistance;

- To determine natural enemies of the pests and their potential for use in biological control; and
- To investigate cultural practices that will limit the pests.

Cassava pest control

Cassava tolerates, and compensates for, pest damage well. The reason is that it has a long growth period (1–2 years). Thus, during the dry season, when the pests in Africa are a problem, the plant grows very slowly because of the water shortage but is able to compensate during the following rainy season.

This ability of cassava to tolerate and compensate for damage caused by insects and mites makes it especially attractive for development of resistant or tolerant varieties because the level of resistance or tolerance need not be too high.

Programs to identify resistance against *Zonocerus*, *Mononychellus*, and *Phenacoccus* are under way and have already yielded some positive results. High levels of resistance have been found for *M. tanajoa* in Zanzibar and Ukuriguru (Shukla 1975) in Tanzania and in IITA. Plants exhibiting resistance have hairy young leaves and good vigour. Also, some resistance to *Phenacoccus* may have been recently identified, as Ezumah et al. (1979) have reported clones on which no mealybug colonies become established. Likewise, for the red spidermite, clones that have reduced susceptibility have been found, although the level of resistance is not yet satisfactory. Finally, clones on which a *Zonocerus* attacked and fed on only the lower leaves have been identified (Leuschner 1977a).

These findings are encouraging; it is now up to breeders to utilize the resistant clones and, if possible, to combine their characteristics with other desirable characters like disease resistance and high-yield potential.

As a supplementary control, it is always desirable to have a complex of natural enemies of each pest to keep it at low levels when other means fail or are not satisfactory. These biological controls already exist in Africa for indigenous pests like the red spidermite and scale insects; however, *Mononychellus* and *Phenacoccus* were introduced to the continent without their natural parasites and predators and, hence, have spread quickly.

A biological control program has been started by IITA, in collaboration with the Commonwealth Institute of Biological Control, for the mealybug and green mite. Predators and parasites will be sampled in South America, identified, and tested for their efficiency.

In general, pest outbreaks and damage to plants are more severe when the plants are under stress from other factors such as drought, poor soil fertility, or poor adaptation to a certain environment. This is certainly the case with spider mites, mealybugs, and scale insects. Observations in Zaire of cassava planted on rich alluvial soils with a high water table compared with that on poor laterite soils with a low water table indicated that the mealybug infestation was much higher on the poor soils. The same applies to the green mite in Tanzania.

Planting time is another factor contributing to the amount of infestation by mealybugs and green mites and the consequent yield losses. Cassava plants younger than 6 months suffer greater yield losses than do older plants with well-established root systems.

Therefore, things that improve plant health, like crop rotation, an increase of organic matter (mulch), use of healthy cuttings, and proper planting time, can significantly reduce yield loss. In the case of mealybug attack on cassava in Zaire, 4 months' delay in planting resulted in 31.2% yield loss compared with none for cassava planted at the beginning of the rainy season. The effects of other agricultural practices, like mixed-cropping systems, have to be investigated thoroughly in Africa. For example, mixed-cropping systems reduce pest incidence, but the reason is not clear. The low population of each crop and the influence of barrier crops are possible factors.

Because of the low unit value of cassava, as well as the socioeconomic constraints to cassava production, insecticides and acaricides should not be used in intensive cassava production. Chemical control should be limited to the treatment of planting materials. Treating the stakes with malathion or kelthane delays infestation by green mites and mealybugs considerably. Reinfestation can only take place by airborne crawlers and mites, and their density is low at the beginning of the dry season (Nwanze 1978). In the case of mealybugs in Zaire, it is only in August–September that crawler density markedly increases.

Sweet-potato pest control

Compared with cassava, sweet potato is a short-lived crop. The time from planting to harvest is 4–6 months. Its economic threshold for pest damage is much lower because it cannot compensate by growing for a longer period. Therefore, pest-control methods must be more efficient. Fortunately, sweet potatoes have only one economically important pest (*C. puncticollis*); therefore, control strategies can focus on it.

As with cassava, host-plant resistance against *Cylas* would be the easiest way to control this insect in Africa because no additional input by the farmers would be required. Unfortunately, breeding for resistance is still in the early stages: 700 cultivars collected from all over the world have been screened but only low levels of resistance have been found, and crossing the clones with each other has only produced clones with moderate levels of resistance. The difference in susceptibility (Fig. 2) between the moderate and susceptible clones (Hahn 1977b; Leuschner 1977a) is not sufficient, especially at the beginning of the dry season. Thus, additional means of control have been studied.

Biological control against the weevil is not promising at present either, because no efficient parasites or predators have been found to combat the pest during the dry season. Nevertheless, one fungus, a *Boveria* sp., has been found to be very efficient under the humid conditions of the wet season, and more investigations are needed.

Various cultural-control methods have been tested, and two factors have been shown to contribute significantly to the control of the weevil: use of clean planting material (cuttings) and reridging of plots. Clean cuttings can be obtained from the end of the vine, but larvae have been found to exist everywhere else on the vine. Reridging is an old method practiced by people in Indonesia. The idea behind it is to keep the growing tuber deeply buried because *Cylas* weevils are not able to penetrate the soil; they depend on cracks and other crevices to reach the tuber.

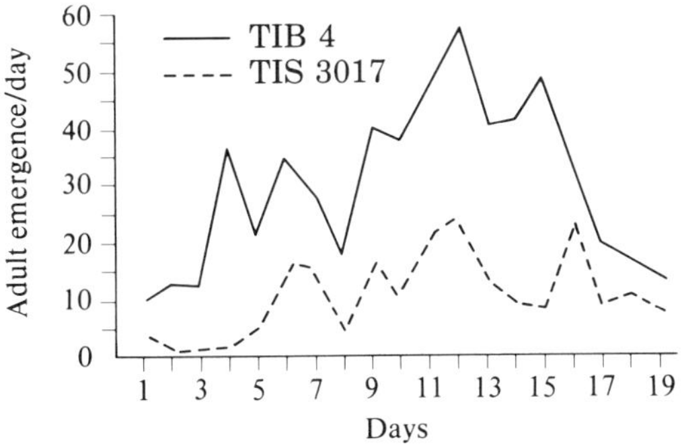

Fig. 2. *Adult weevils emerging from a susceptible (TIB 4) and a field-resistant sweet-potato clone (TIS 3017) under laboratory conditions.*

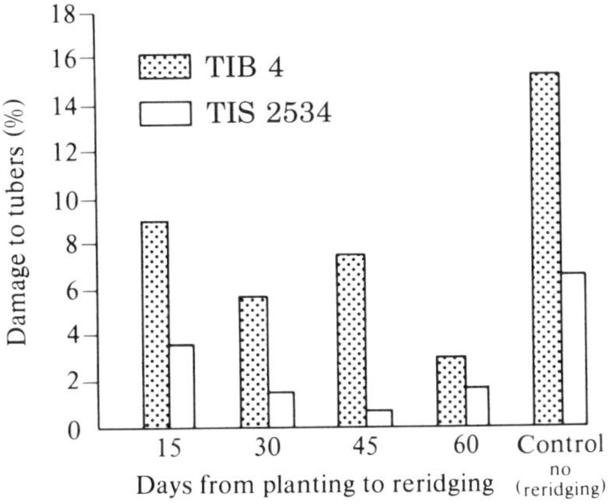

Fig. 3. *Wet season reridging trial (sweet potato).*

In two trials conducted in the beginning and end of the rainy season, the best sweet-potato variety in terms of weevil resistance was compared with a susceptible one. During the first 2 months of the growing season, every 15 days five blocks of the trial were reridged. After 30–45 days, the combination of low-level resistance and reridging resulted in a very low amount of damage to tubers (Fig. 3). This shows that, without insecticides, it is possible to reduce the level of weevil infestation (Leuschner 1977a).

The problem of weevil damage during storage of sweet potatoes remains. In storage, even a small weevil population can heavily damage the harvest within 5 weeks. In this case, insecticides must be used, and actellic and synthetic pyrethroids have proved satisfactory when applied once at the beginning of storage.

Quarantine

Root crops have one thing in common: they are vegetatively propagated, the planting material serving as transportation for pests like mites, mealybugs, and scale insects. In Africa, diseases or pests that occur in one country are spread over the whole continent within a few years because of the increasing movement of people. Within the last 10 years, cassava alone has faced two new problems imported: mealybugs and green mites.

It is, therefore, essential that all governments exercise controls and quarantine on planting materials. New varieties of root crops should be imported only in seed form. These seeds should pass through a plant-quarantine station, or, if this is not possible, they should be planted in isolation.

Cassava green mite: its distribution and possible control

Z.M. Nyiira[1]

The cassava green mite (*Mononychellus tanajoa*), not known in Africa prior to 1971, is now distributed in East, Central, and West Africa. Investigations to date have covered its identification, biology, ecology, and control. Recently, it has been claimed that *M. tanajoa* is a synonym for *Eotetranychus caribbeanae*, which attacks cassava in the Caribbean and Latin America and is known to exist in various African countries. Findings so far indicate that the green mite reproduces rapidly; that physical factors affect population and reproduction of the mite; that the pest is airborne and its dispersal across borders is not possible to control; that biological control should be part of an integrated approach; that some cassava cultivars are tolerant or resistant to mite damage; that quick-maturing varieties are subjected to fewer attacks; that plants subjected to stress, e.g., from lack of nutrients or water or from stiff weed competition, during the first 2–3 months of development suffer particularly heavy damage from mite infestation; and that the pest apparently has a 5-year, long-term population cycle as opposed to an intraannual population trend. Another finding is that an *Entomophthora* fungus attacks and kills green mites, but its application in biological control programs is limited because it kills predatory mites and is only effective during the rainy season when the mite population is already greatly reduced.

La teigne du manioc (*Mononychellus tanajoa*), inconnue en Afrique avant 1971, se retrouve maintenant dans l'est, le centre et l'ouest du continent. Les recherches ont porté jusqu'à présent sur son identification, ses caractéristiques biologiques et écologiques et les façons de la neutraliser. On a récemment suggéré que *M. tanajoa* serait identique à *Eotetranychus caribbeanae*, qui s'attaque au manioc dans les Antilles et en Amérique latine et dont l'existence est connue dans plusieurs pays d'Afrique. Jusqu'ici, les découvertes indiquent que la teigne se reproduit rapidement, que des facteurs physiques influencent les populations et la reproduction, que c'est un insecte volant et qu'on ne peut donc l'empêcher de traverser les frontières, qu'une approche intégrée devrait comprendre la lutte biologique, que certains cultivars de manioc ont une tolérance ou une résistance à la teigne, que les variétés à maturité hâtive sont moins sensibles, que les plantes soumises à des contraintes, comme le manque d'éléments nutritifs ou d'eau ou la concurrence serrée des mauvaises herbes, au cours des deux ou trois premiers mois de leur développement, souffrent le plus des ravages de la teigne et que les fluctuations de sa population suivent un cycle de 5 ans plutôt qu'un sommet interannuel. On a aussi découvert que le champignon *Entomophthora* attaque et tue la teigne; il n'a toutefois qu'une utilité limitée pour les programmes de lutte biologique puisqu'il tue les acariens prédateurs et n'est efficace que pendant la saison des pluies, alors que le nombre de teignes a déjà beaucoup diminué.

Cassava production in Africa is low, largely because of poor agronomic practices and pests and diseases. Until the late 1960s, the major constraints were cassava mosaic disease, grasshoppers, and, to a certain extent, locusts. During the last decade, the cassava green mite, cassava bacterial blight, and the mealybug have appeared on the continent and have rapidly become major problems. The green mite (*Mononychellus tanajoa*) is now widespread and is considered to be the most destructive pest of cassava in Africa. It not only feeds upon the plant but also contributes to reduced root size, poor root quality, and late root formation.

In 1971, the first outbreak of green mite was reported in Uganda. Since then, the mite

[1] Kawanda Research Station, Ministry of Agriculture, P.O. Box 7065, Kampala, Uganda.

has been reported in all the countries of the Congo basin except Zambia and Malawi as well as in Sudan, Kenya, and Tanzania. The exact spread of the mite is not documented because of a lack of coordination of survey reports, but it is suspected that the pest has now spread into more countries in Central Africa and possibly into West Africa.

Tuttle et al. (1977) consider *M. tanajoa* to be a synonym for *Eotetranychus caribbeanae*. If this is true, then the distribution of the green mite is wider still, based on specimens of *E. caribbeanae* that I collected on cassava from Malawi and Benin Province of Nigeria in 1977, and from Angola and Zambia in 1978. However, Flechtman (personal communication) still maintains that *M. tanajoa* is distinct from *E. caribbeanae*.

Work to date

Investigative work on the green mite in Africa started just after the first reported outbreak. A survey of its distribution and spread produced important information about the rate of dispersal of this and other species of mites that infest cassava. Dimethoate, chlorobenzilate, and kelthane were found effective against the mites and were tentatively recommended for use by farmers. However, the trials that indicated the effectiveness of these chemicals also indicated that their use in Africa was not feasible because of their cost and their adverse effects on the predators of the green mite, in particular, *Oligota minuta*, *Stethorus* spp., and *Typhlodromus* spp.

During 1973–76, research emphasis was placed on monitoring the mite's spread in Africa, surveying the neotropics for the mite and its natural enemies, and studying its biology, ecology, and control through agronomic practices, biocontrol, and building-in host resistance. The results of research conducted in various countries indicated that:

- The green mite has a high reproductive rate, a short egg-adult duration (11–13 days), and many generations over the cassava crop;
- Physical factors affect the population and reproduction of the mite;
- Dispersal across borders is impossible to control;
- Biological control agents are effective enough to merit application against the mites but are not sufficient alone because the multiplication rate of the mites is higher than that of its predators;
- Some cassava cultivars are more tolerant to mite damage than are others; and
- Quick-maturing varieties are subjected to reduced numbers of challenges.

These findings suggest that integrated control of the green mite is possible and appropriate in the long run. It would, of necessity, combine tolerant varieties planted at an optimum time with biological control efforts.

Local cultivars and those obtained from IITA have been tested for resistance/tolerance in Uganda and Tanzania since 1976. The results from the various trials have been reported by different workers, including Carpenter (personal communication) in Zanzibar. Other trials, not reported, have shown promise for release, although the yield parameters are rather low. Work in Brazil has also indicated the presence of host resistance against the green mite in cassava.

The economic importance of the mite has been investigated in simulated trials and field conditions. My work (1976) indicated a loss in yield of fresh roots of up to 40%. In South America, trials have indicated yield losses of up to 20% under field conditions. The difference is attributable to physical factors that suppress populations and damage. More recent evaluations in Uganda of yield losses under field conditions have ranged between 17% and 33%, depending on whether cassava was planted on newly opened or continuously cultivated land. Furthermore, indications are that cassava grown on poor soil in dry areas or cassava growing under stiff weed competition suffers increased damage and yield losses.

Annual population trends have been found to be associated more with the absence of leaves on plants than with the presence or absence of rain, although, generally, the population of mites is lower during the wet season. A 5-year population cycle has been observed in Uganda: damage from the mite was devastating in 1972–73 but subsided to a tolerable level until 1977–78 when it again became excessive. The factors influencing this long-term population trend are not known, and records are being kept to confirm this observation.

Biological control

During 1979, I observed that aphids and

Table 1. Percentage mortality of *M. tanajoa*, *T. bimaculatus*, and *Phytoseiulus* spp. following application of *Entomophthora* suspension on cassava leaves artificially infested with mites in the laboratory.

Days after treatment	*M. tanajoa*	*T. bimaculatus*	*Phytoseiulus* spp.
7	23	19	27
10	64	69	70
20	91	93	87

mites on citrus leaves had been attacked and killed by a fungus. Similarly, aphids on *Brassica* were found dead from fungal infection. Samples of the infected aphids were collected and ground; the resulting powder was mixed with water, and the suspension was sprayed on cassava leaves infested with green mites. After 10 days, more than 50% of the mites were dead through fungal infection. A look-see trial was set up as a follow-up, and the results (Table 1) showed that up to 91% mortality could be achieved 20 days after application of the suspension.

Although factors like the age of test mites, the age of the fungal culture, temperature, and restriction of test mites were not controlled in this trial, the results demonstrated the effectiveness of the fungal suspension against not only the green mite but also *Tetranychus bimaculatus*, another pest of cassava. Unfortunately, the fungus also killed the beneficial predator, *Phytoseiulus*. The fungus was identified as *Entomophthora ?thaxteriana* Petch.

In practice, this fungus cannot be used in biological control because the humid conditions that favour its prevalence discourage mite-population development anyway.

Discussion

The cassava green mite presents one of the major threats to improved cassava production, and there is so far no effective recommended control for it. In fact, the information on its biology and ecology and investigations conducted to develop integrated control measures are, at present, poorly coordinated even though there is sufficient information on which to base an integrated control program of a pilot nature.

Currently, the most feasible, practical approach is to use early-maturing, resistant varieties and to plant them when they will not be seriously challenged during their early development. Biological control alone will probably not be sufficient but should be integrated with strategic agronomic practices. Chemical control should be discouraged because of its effect on predators as well as its cost.

Not enough is known about appropriate planting date, and the development of early-maturing, resistant cultivars is time-consuming. However, work is under way and, in my opinion, would benefit markedly if:

- A coordinating body were established to make an inventory of research into the cassava green mite on the continent;
- Cassava varieties, resistant to *M. tanajoa*, were documented, multiplied, and released to farmers while being investigated in trials of optimum planting dates;
- A central multiplication unit were established to ensure safe distribution of clean, resistant stock; and
- Pilot projects were introduced in selected countries in collaboration with IITA. The aim would be to speed up research on the green mite and to confirm and demonstrate results.

The biological control program initiated in Kenya by the Commonwealth Institute of Biological Control has been temporarily stopped, pending establishment of facilities. Predators released in Kenya during 1978 have not been recovered, but the *Oligota* species released and recovered in Uganda indicate that the buildup of the predator is much slower than that of the mite population, especially when there is a prolonged dry season.

Biological control of the mite in Uganda has been complicated by the prevalence of cassava bacterial blight, which increases in severity during the wet months when the population of the mites has subsided.

A collaborative program in biological control should incorporate Zaire, Uganda, Rwanda, Tanzania, and probably Burundi on the one hand and Uganda and Kenya on the other. It should be part of an integrated effort that relies primarily on the development of resistant, quick-maturing varieties planted at a time that allows formation of roots before the buildup of the mite population.

Biological control of cassava mealybug and cassava green mite: front-line release strategy

K.M. Lema and H.R. Herren[1]

Biological control is one of the most promising solutions to the problems caused by the cassava mealybug (*Phenacoccus manihoti*) and the cassava green mite (*Mononychellus tanajoa*). At present, however, there are not effective natural enemies in Africa; they must be identified in, and imported from, South America where the two pests are indigenous. This is the task being undertaken by entomologists at IITA, and some progress has already been made. One species of predator imported during the first phase of the efforts is being successfully mass-produced and will soon be released against the mealybug. The release may be done at the front as well as inside infested areas.

La lutte biologique est l'une des solutions les plus prometteuses au problème de la perte de manioc due à la cochenille (*Phenacoccus manihoti*) et à la teigne (*Mononychellus tanajoa*). Il n'y a toutefois actuellement aucun ennemi naturel efficace en Afrique et il faut les identifier et les importer d'Amérique du Sud d'où ils sont originaires. Les entomologistes de l'IITA s'occupent de cette question et ont déjà réalisé des progrès. On a réussi la production massive d'un des prédateurs importés au cours de la première phase des travaux et on va bientôt le lancer contre la cochenille, autour et à l'intérieur des régions infestées.

Cassava in Africa was free from major arthropod pests before 1970. Since then, however, the plant has been seriously affected in many parts of the continent by two pests, the cassava mealybug (*Phenacoccus manihoti*) and the cassava green mite (*Mononychellus tanajoa*), both introduced to Africa from South America.

The cassava mealybug was first reported in Zaire in 1973 (Hahn and Williams 1973) but was described and named only in 1977 (Matile-Ferrero 1977). Since its discovery, the pest has spread to several countries, and its biology has been studied by Nwanze (1978) and Fabres and Boussiengue (1980).

The cassava green mite was introduced into Africa through Uganda and was first found near Kampala in November 1972. Its biology, distribution, and ecology in Africa were studied by many workers, including Nyiira (1976, 1977, 1979).

These two pests cause large losses in leaf and root yield and, thus, are a threat to cassava production in several African countries. Because of the importance of cassava as a staple food in Africa, urgent measures should be taken to contain these pests.

Good cultural practices (manipulation of planting time, intercropping, use of pest-free planting materials) must be a part of any pest-control program. However, alone, they are not sufficient to control the mealybug and green mite. Chemical control may be helpful, but its costs and side-effects make it unsuitable for use in Africa except in the treatment of planting materials. Even developing host-plant resistance, which is a lasting solution and an excellent approach, has its drawbacks: it takes years, and the resulting varieties may perform well in one area but not in another. The most promising and permanent solution in the short term is biological control.

The term biological control may sound new, but the idea of using insects to control other insects is ancient. In fact, as far back as the 3rd century, the Chinese placed the nests of a predatory ant (*Oecophylla smaragdina*) in citrus orchards to control insect pests (Needham 1954, cited by Konishi and Ito 1973).

Biological control has already proved successful against the mealybug. Successes

[1] International Institute of Tropical Agriculture (IITA), P.M.B. 5320, Ibadan, Nigeria.

against the citrus mealybug (*Planococcus citri*) in California, USA, and the coffee mealybug (*P. kenyae*) in Kenya are two outstanding examples.

Tetranychid mites have also been successfully controlled through the use of natural enemies both in the greenhouse and in the field (Oatman and McMurtry 1966; Hussey and Scopes 1977). In fact, the beetle *Oligota minuta* has been colonized (released) in Kenya to control the cassava green mite, and the prospects of success are quite high (Girling et al. 1979).

In Africa, the two pests lack effective natural enemies (Matile-Ferrero 1977; Nwanze 1978; Girling et al. 1979). Local natural enemies, most of them predators, have not evolved with these two pests but have "switched over" from other similar pests. As stated by Doutt and DeBach (1964), such natural-enemy complexes generally do not regulate newly introduced pests. Therefore, effective biological control agents must be imported from South America, mass-produced, and released into infested areas of Africa.

An international network of scientists (in Africa and Latin America including specialists from the Commonwealth Institute of Biological Control, CIBC) has been developed to deal with the exploration, collection, rearing, and distribution of the natural enemies. At IITA, the emphasis is being put, first, on the biological control of the cassava mealybug.

The biological control program includes:

- Exploration for, and importation of, efficient natural enemies (in cooperation with CIBC) along with biological, ecological, and behavioural studies of imported natural enemies, including the development of good methods for their mass production and release;
- Biological and ecological studies of both the cream (in cooperation with Centro Internacional de Agricultura Tropical, CIAT) and the pink form of *P. manihoti*; and
- Release and assessment of the effectiveness of the natural enemies, as well as economic analysis of the biological control program.

Several species of parasitoids and predators collected from South America by CIBC specialists were, after quarantine screening, brought to Nigeria (IITA) where they are being studied. They were collected from the cream-coloured form of *P. manihoti* (the bisexual form found in South America). Parasitoids do not successfully reproduce on the pink form (parthenogenetic) of the mealybug found in Africa. Such difficulties have been well documented in the past when parasitoids from one pest species were erroneously imported against another pest species or strain, as in the cases of the coffee mealybug in Kenya (DeBach 1974) and the grape mealybug (*Pseudococcus maritimus*) in South Africa (Bartlett 1978).

Predators are generally less specific than parasitoids in their selection of host species. One of the species imported from South America is being successfully mass-reared on the pink form and will soon be released in the field.

Front-line release strategy

In classical biological control, mass-produced natural enemies have generally been released inside infested areas. In a front-line release strategy, the releases are made at the front. This idea is not new; Girling et al. (1979) suggested it for the control of the cassava green mite. The strategy is most useful against introduced pests that progressively cover new areas of an invaded country.

When an exotic pest is first reported in a country, a survey should be made to determine its distribution or spread within that country. From the point of entry into the country, the pest spreads quickly, but it takes time for the pest to invade the whole country, if, indeed, the climatic and topographic conditions as well as availability of the host plant allow the pest to do so. Thus, a line (front line) can be drawn between infested and noninfested regions.

In the early stages of infestation when most of a country or region has not been infested, we believe that great emphasis should be put on protecting the pest-free areas and that parasitoids and predators should be released along the front line(s) to reduce or delay the spread of the pest. Complementary releases may be made inside infested areas either after or during releases at the front. Such a procedure would both reduce pest populations within infested areas and protect pest-free regions.

The mealybug problem and its control

T.P. Singh[1]

Although only recently introduced to Africa, the mealybug (*Phenacoccus manihoti*) has been reported to cause yield reductions of 57–85%. It is a dry-season pest and proliferates rapidly at temperatures between 27° and 29°C. It multiplies parthenogenetically, and a single female is sufficient to cause infestation. In the field, the pest is spread by wind and, over large distances, through the movement of infested planting material. Control of the pest is being investigated through both short-term (cultural and chemical) and long-term (biological and resistance-breeding) measures, and some progress has been made. For instance, studies on cultural control have revealed that early planting and mulching of the cassava crop can reduce the damage from the pest and that chemical treatment of planting material is desirable, although chemical control does not seem feasible in areas, such as Zaire, where the cassava leaves are eaten as a vegetable. Parasites and predators found in Zaire have been of little help in controlling the mealybug; therefore, biological control agents will have to come from outside the country. Resistance to mealybug has been identified both in wild and in cultivated cassava types. Some problems, however, have been encountered in the use of this resistance.

Bien qu'arrivée récemment en Afrique, la cochenille (*Phenacoccus manihoti*) a fait diminuer les rendements de 57 % à 85 %. Il s'agit d'un insecte de saison sèche qui prolifère rapidement à des températures de 27 à 29 °C. Il se multiplie par parthénogénèse et il suffit d'une femelle pour provoquer l'infestation. L'insecte se propage sur le terrain par le vent, et sur de grandes distances, par le transport de plants infestés. On étudie les mesures de lutte à court terme (agricoles et chimiques) et à long terme (biologiques et sélection en fonction de la résistance) et on a réalisé des progrès. Par exemple, les études de lutte agricole ont révélé que le manioc planté et paillé tôt résiste mieux et que le traitement chimique des plants était souhaitable bien que la lutte chimique ne soit pas toujours possible dans les pays comme le Zaïre, où l'on consomme les feuilles de manioc. Les parasites et les prédateurs trouvés au Zaïre n'ont que peu d'effet sur la cochenille et il va falloir importer les agents de lutte biologique. On a identifié la résistance à la cochenille chez des types de manioc sauvage et cultivé, mais on éprouve quelques difficultés à en tirer parti.

Mealybug (*Phenacoccus manihoti*) is one of the two pests of cassava that have become serious constraints to production in Africa during the past decade. Since it was first reported in Zaire by Hahn and Williams (1973), the pest has spread far and wide in Central and West Africa. It has been reported in Zaire, Gabon, Angola, the Congo, Nigeria, and Senegal (Hahn et al. 1978). Serious attack by the mealybug can cause a root-yield reduction of 54–85% (PRONAM 1979).

Because of the serious implications of mealybug infestation, a concerted research effort was mounted by scientists in Programme national manioc (PRONAM) to find measures to control this menace. Initially, a survey was conducted to determine the pattern of pest distribution and its severity. Studies were also initiated on the biology of the pest and the factors influencing pest populations. They revealed that the mealybug is a pest during the dry season and that its population reaches a peak when the temperature is high (27°–29°C) and relative humidity is about 65–70% (Leuschner 1977b; Nwanze 1978). The mealybug has been reported to be particularly damaging to plants growing in poor soils (Nwanze 1978). Biological studies have revealed that *P. manihoti* is female and that it multiplies parthenogenetically. A single female lays an average 440 eggs, which

[1] Programme national manioc (PRONAM), B.P. 11635, Kinshasa, Zaire.

hatch in 8 days. Crawlers, which take 12 days to reach the adult stage (fourth instar), emerge and move to the upper leaves and plant tip where they are picked up by the wind and are spread throughout the field. Another means of spread is through the movement of vegetative planting material. A single gravid female is all that is necessary to initiate infestation (Nwanze 1978).

The majority (60%) of the insects are found on the under surface of the leaf, and, of these, about half (52%) are oriented on the leaf midrib, and most of the remainder are found on secondary veins. The first symptom of mealybug damage is clustering and curling of top new leaves and is called "bunchy top." The shortening of internodes follows, and this may be caused by the introduction of some kind of toxin into the plant. The final stage of pest damage is referred to as "candlestick" (Leuschner 1977b) — death of the growing point, which is covered with a white mass. Studies on root-yield losses conducted by PRONAM (1977) indicated reductions in numbers, length, and girth of roots (Table 1).

For control of the mealybug, short- and long-term approaches were envisaged. The short-term approach includes cultural and chemical means, whereas the long-term approach includes biological control and resistance breeding.

Cultural control

Because the pest is most serious on young plants subjected to water stress in the dry season, experiments were conducted to determine whether maintaining soil humidity at a high level by mulching and planting cassava early in the rainy season would help reduce mealybug damage. Results clearly showed that mulching decreased the degree of pest damage and the number of plants attacked, irrespective of method of planting (Table 2) (PRONAM 1977).

Plant age at infestation has a significant effect on root yield, especially when the infestation starts before the plants are 8 months old (PRONAM 1978). Thus, it is recommended that the crop be planted as early as possible (November–December) and that mulch be used to conserve soil moisture.

Chemical control

Chemical means of controlling the pest are considered to be much quicker and more efficient than other means, but, in Zaire, where cassava leaves are frequently eaten by the population, the spraying of cassava for mealybug protection cannot be recommended. Other factors militating against the use of

Table 1. Effect of mealybug attack on the root yield (t/ha) of two cassava varieties grown on two soils.

	02864 grown on:		
	Valley soil	Heavily eroded soil	Mpelo-Longi grown on lateritic soil
Noninfested	20.5	16.4	9.6
Mealybug infested	9.4	3.2	1.5
Yield reduction (%)	54.4	80.8	84.4

Table 2. Effect of mulch on mealybug incidence on two varieties of cassava.

		02864		Mpelo-Longi	
Treatment	Subtreatment	Pest score	Percentage attack	Pest score	Percentage attack
Ridges	Mulch	1.5	35.9	1.0	12.1
	No mulch	2.3	49.2	2.5	54.7
Flat	Mulch	1.3	32.1	1.3	16.6
	No mulch	2.3	61.0	1.9	23.6
Ridges, flat	Mulch	1.4	34.0	1.2	14.4
	No mulch	2.3	55.1	2.2	39.2

chemical control are: the high cost of the insecticides; their unavailability; and the farmers' almost complete ignorance of the safe use of chemicals and spraying equipment. However, some experiments were conducted with chemicals in treatments of cassava planting material, and the results clearly showed that preplant dipping of cassava stakes in a solution of dimethoate (500 g a.i.) (5 ml/L of water) helped to free the stakes from crawlers, although it did not give any protection against later field infestation (Nwanze 1978). The establishment of stakes after insecticide treatment was good (PRONAM 1978).

Biological control

Biological control is a new approach for the control of cassava mealybug in Africa. Preliminary work initiated in PRONAM for the biological control of mealybug has been directed toward identifying parasites and predators that are available locally and determining their efficiency in checking mealybug populations. Findings were that the most predominant predators and parasites are coccinellids and hymenoptera and that biological control of mealybugs by local species is taking place but to a limited degree. Some parasites and predators have been introduced through the Commonwealth Institute of Biological Control (CIBC), releases being made in three locations, but no recovery was possible (PRONAM 1978). The International Institute of Tropical Agriculture (IITA) has recently initiated a concerted research effort to control mealybug through the introduction of exotic parasites and predators.

Resistance breeding

A considerable amount of genetic variability in the form of seed was introduced into Zaire from IITA and other sources for screening against the mealybug. Under artificial conditions, three wild cassava plants resistant to mealybug were identified (PRONAM 1978). These plants do not produce edible roots, however, and the normal procedure for multiplying mealybug-resistant plants failed because the stem cuttings did not root to support the sprouting buds. The plants were incorporated in a hybridization program to transfer their gene for resistance into locally adapted varieties. Some success was obtained but the seeds that resulted showed poor germination under field conditions. Grafting the mealybug-resistant plant as a scion on the stock of cultivated cassava has also been tried, with a success rate of about 10–15%.

Efforts to find an additional source of resistance, especially in cultivated cassava, are continuing, and some promising results have been obtained from another batch of cassava seed introduced from IITA. The first negative screening of 6504 individual plants under artificial infestation resulted in the identification of 123 plants with relatively little mealybug damage (PRONAM 1979). The number of plants for the scores 1–5 were 42, 1939, 1916, 1850, and 757, respectively. Five cuttings from each of the selected plants were planted in May 1980 so that their level of resistance could be confirmed. Each plant in a row was infested artificially in August, and the clones were screened in October before the rains. Of 104 clones, 34 (32.7%) showed pest scores of between 1 and 2, whereas the local variety (02864), planted as a check, always showed a score of 4. Of these 34 clones, 12 showed a score of 1–1.5. Thus, the potential for finding a source of resistance to mealybug is good.

The efforts made so far in PRONAM on the control of mealybugs suggest that a long-lasting solution to the menace is in sight and can be achieved through an integrated control approach.

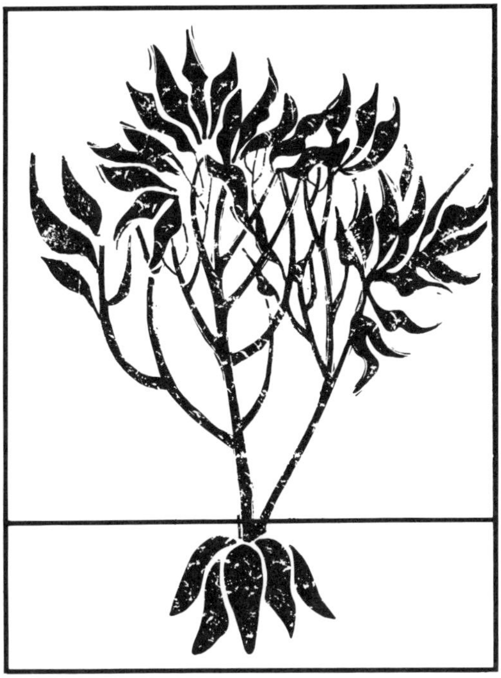

Agronomy

Sweet potatoes — variety TIB 1 — being raised for distribution to farmers.

Economics of research and development of root and tuber crops in Zanzibar, Tanzania

A.J. Carpenter[1]

A total investment of about $100 000, from IDRC, IITA, the Canadian University Service Overseas (CUSO), FAO, and the Ministry of Agriculture, Zanzibar, resulted in the development, to the first stage of rapid multiplication, of several pest- and disease-resistant cassava varieties. The best of these promise to produce an extra 2 t of cassava on each 0.4-ha farm. At current local prices (US$ 0.15/kg on farm), the extra production is worth $300 to the farmer, or about $8 million if 50% of the current cassava-growing area in the country were planted with the improved varieties. Although this crop-value increase would be a good rate of return on an investment of as much as $20 million, estimates suggest that an investment of only $700 000 (at 1978 prices) will be needed for the extension program. It is concluded that research and development in root-crop improvement is highly attractive in economic terms, provided that market and family demand is maintained.

Des injections de crédit totalisant environ 100 000 $ du CRDI, de l'IITA, du Service universitaire canadien outre-mer (SUCO), de la FAO et du ministère de l'Agriculture de Zanzibar ont permis le développement, jusqu'à l'étape de multiplication rapide, de plusieurs variétés de manioc résistantes aux insectes et aux maladies, dont la meilleure pourra sans doute produire deux tonnes de manioc de plus par ferme de 0,4 ha. Aux prix locaux actuels (0,15 $US/kg à la ferme), le surplus rapporterait 300 $ à chaque fermier, soit 8 millions de dollars si la nouvelle variété était semée sur la moitié des plantations actuelles de manioc au pays. Bien que la valeur accrue de la récolte puisse rentabiliser un investissement de 20 millions, on prévoit qu'il suffira de 700 000 $ (en $ de 1978) pour mener à bien le programme d'extension. On conclut que la recherche et le développement en matière d'amélioration des récoltes offre des perspectives économiques intéressantes pour autant que la demande commerciale et familiale se maintienne.

Compared with industrial research and development (R&D) expenditures, investments in R&D for tropical agriculture usually appear small; however, the reports from well-designed crop-improvement projects almost always indicate that large increases in production and profits are possible in most tropical farming systems. The cassava-based farming systems in the Zanzibar islands are no exception, and efforts have been made to improve them. (An appraisal of the stream of costs and benefits is also attempted.)

Cassava is grown mainly on well-drained sandy and red soils on ridges that tend to run northwest to southeast across the islands. The soils are mainly devoted to tree crops, almost exclusively cloves in Pemba and cloves, coconuts, and fruit trees such as mango and durian in Zanzibar. Cassava is grown on the light, poor soils where tree cover is less dense, particularly under mature coconuts (the better and more densely treed soils support an understory of food crops such as banana and cocoyam). Cassava is also grown in pockets of deeper soil on the reefal limestone or coral rag areas that predominate on the eastern halves of both big islands. Stakes are planted on ridges (about 1×10^4 plants/ha). Farmers normally plant a selection of other crops on the ridge: sweet potatoes are almost universal, and corn, tomatoes, radishes, yams, and cocoyams are frequent. What adverse effect this intercropping has on cassava yields is unknown, but it certainly produces a useful flow of food while the cassava is developing.

[1] Food and Agriculture Organization of the United Nations (FAO), P.O. Box 159, Zanzibar, Tanzania.

Cassava is usually planted after the main rain in June and July. Some quick-maturing, low-yielding varieties are harvested from 6 months after planting, but most of the crop is harvested from 12 months onward.

Yields are universally low, ranging from 2 t/ha to 4 t/ha but are complemented by the yields from the intercrops, including the trees, as they are only a part of the whole production system.

Labour inputs are estimated at 600 work-days/ha, two-thirds of which can be attributed to the cassava crop. Market prices of cassava have risen rapidly, from about US$ 0.15/kg in 1975 to US$ 0.30–0.60/kg in 1980. Taking "on-farm" value as half the lower figure, I estimate that a farmer produces about $450 worth of cassava per hectare or about $1.00 return per day's work. This compares unfavourably with clove harvesting and yam and cocoyam production, but it is similar to, or better than, pulse production and unimproved rainfed rice production. Because the average cassava farm is about 0.4 ha, the potential value of cassava per family is about $150 (plus the intercrop values). Along with cash, the cassava forms a readily available source of reserve energy. The low yields are mainly a reflection of the intrinsically low-yielding character of the local clones, a fact that has been established by comparisons of local cultivars grown side by side with improved material. Widespread damage by cassava mosaic disease (CMD) and, more recently, cassava green mite (CGM) also contributes to the low yields. Other contributing factors are the practice of intercropping; the poor soils, which are farmed without rotation; and neglect of the crops, with heavy weed growth often being permitted after the intercrops are harvested. Also, there is a strong preference among the population for sweet, succulent varieties, and, although these taste characters are not necessarily correlated with low yield, conservation and propagation of high-yielding but fibrous clones would be unusual at the family level.

The diet of farm families in Zanzibar is unusually varied and includes legumes, coconuts, cocoyams, bananas, sweet potatoes, *Amaranthus* spinach, fruits, and usually some milk, poultry, and fish. Rice forms the staple and preferred food, however, and recent efforts to improve rice production have dominated the agriculture R&D and investment scene. Nevertheless, the high potential for improved yields in cassava has not been ignored entirely: in 1976, for example, some seed material supplied by IITA was tested. Further efforts were stimulated by the invasion of CGM early in 1977 and the realization that substantial varietal resistance to this pest, combined with CMD resistance and high yield, was available in the material under study. A small program was financed by IDRC to cover the introduction of more seed material, the training of field staff at IITA, capital provision for transport and equipment, and technical assistance.

The first results of this program have now been in rapid multiplication for almost a year, and Kilimo is poised to distribute better materials to the island's farmers. Several clones in multilocation trials yielded 5–10 t/ha under normal farm-management conditions, their roots having a flavour as good as or better than the preferred local types. The potential extra value of this material is substantial. From the estimated 25 000 ha of cassava grown annually, the extra yield from 50% adoption of the new varieties would be 50 000 t/year valued (on farm) at almost $8 million. Farm families adopting the new variety would average an extra 1.5–2 t of cassava from a 0.4-ha holding. The cash-equivalent earnings would rise to $480, which is a return of $3/day, or about the same as average clove-picking earnings or improved rainfed rice production returns.

What was the cost of the R&D to produce this potential production and income increase? Including the IDRC grant, technical assistance, Ministry costs, specialist travel, and an allowance for IITA's investment, no more than $100 000 was spent on the whole program up to the first stage of multiplication. Thus, the return could be conservatively set at $300 on each dollar invested, with the assumption that the R&D will continue at $25 000/year as a permanent program. It is difficult to imagine another (legal) activity that could match this rate of return.

To get a better idea of the actual cost of realizing the $8 million, one must estimate the extension requirements for getting the new material to 50% of the farmers. These costs were estimated by Virginia Price in late 1979. Capital and 1st-year running costs were estimated at $400 000, with subsequent recurrent costs of about $140 000. In other words, the total R&D and extension costs would be about $200 000/year, a return of

40 : 1 once 50% of the farmers were growing the new material. On a continuing basis, one increment of production worth $10/ha annually would pay for a permanent service (R&D plus extension). This increment represents 70 kg of cassava per hectare, or the yield to be expected from 12 good plants.

In investment banking, a rule of thumb is that, if the annual extra value of the crop equals or exceeds one-third of the total investment, the project will give an excellent rate of return. On this basis, the extra value potentially available from the cassava-improvement program would support a total investment of $20–$30 million, although most banks would need considerable assurance as to price stability and elasticity of demand for cassava before advancing such a sum.

This analysis of the costs and benefits presents the cash equivalent for improved cassava production based on the current activities of farmers and technical staff. But there are many other options in the future. For example, a farmer may increase his or her family energy-food consumption or sell or exchange the surplus. Alternatively, the cassava production could be maintained, some of the area currently cropped with cassava being made available to other crops or the labour input being concentrated on other profit-making or cultural activities.

From the R&D and extension viewpoint, the sky is the limit. The program has already produced bitter varieties that yield more than 20 t/ha. These probably have an immediate outlet as an energy source for the Zanzibar feed mill and might be grown either by the Ministry or by private commercial growers. There seems little doubt that hybridization can quickly produce higher-yielding clones with the locally popular sweet, succulent quality. A fairly large program will be needed to achieve this quickly. One of the most attractive options for R&D is to expand research into the intercrop species grown in the cassava system. Work on sweet potato has already started and will soon get a boost from the introduction of tissue-culture clones of proven yield from IITA. Maize improvements have also been demonstrated. A maize–cowpea–tomato intercrop (all improved varieties) would probably give a striking return in combination with the high-yielding cassava. The use of *Leucaena* as N-fixing mulch and as fodder for cows in a zero-tillage system also looks promising. Yams and lima beans can be grown on the *Leucaena* stems after they have been cut. The export, by sail boat, of mosaic-free, high-yielding cassava wood for extension distribution as cuttings in mainland Tanzania appears to be an attractive way of capitalizing on the advances made by the Kilimo-IDRC-FAO-IITA-Zanzibar R&D effort in root crops.

Agronomic research on cassava cultivation in Rwanda

J. Mulindangabo[1]

Cassava was introduced into Rwanda in the 1930s and quickly became a staple food for the people. Research on the crop in the country has been under way since the 1940s but received new impetus in 1975. It is now a focus of the Institut des sciences agronomiques du Rwanda (ISAR), which is cooperating with the International Institute of Tropical Agriculture (IITA) in an attempt to reduce the production constraints. ISAR has adopted a 6-year selection procedure, based on IITA standards, for its long-term research program and, on the short term, is involved in assessing the local varieties in its collection so that it can provide assistance to rural farmers.

Le manioc, introduit au Rwanda dans les années 1930, est rapidement devenu une denrée alimentaire de base. Les recherches sur la production se font depuis les années 1940 mais elles ont reçu un nouvel élan en 1975. Elles sont l'une des préoccupations principales de l'ISAR, l'Institut des sciences agronomiques du Rwanda, qui coopère avec l'IITA, l'Institut international pour l'agriculture tropicale, pour tenter de réduire les facteurs limitants de la production. L'ISAR a adopté un programme de sélection de six ans inspiré des normes de l'IITA à titre de projet à long terme, et évalue à court terme les variétés locales de sa collection pour aider les exploitants agricoles.

Cassava (*Manihot esculenta*) was introduced into Rwanda in 1936 by Belgian agronomists to enable the country's inhabitants to survive the famine raging at the time. Since then, it has played an increasingly important role as a staple food for the people. It is drought tolerant, grows well in poor soils, and gives high yields in areas that are marginal for other crops. It can be grown practically everywhere in Rwanda below 1800 m, although the highest yields are obtained at elevations between 1300 m and 1600 m.

The country can be divided into four major regions, according to cassava-production potential:

- Extremely high-yield region: Bugesera and Imbo;
- High-yield region: Mayaga, eastern plateau, and the eastern savanna;
- Moderate-yield region: Impara, central plateau, and the granitic ridge; and
- Poor-yield or no-yield region: shores of Lake Kivu, volcanic soils, the Zaire–Nile divide, and the Buberuka region. In this region, cassava is being replaced by potatoes, which are better adapted to the prevailing conditions.

Past research

The first cassava trials in the country were carried out at the Institut des sciences agronomiques du Rwanda (ISAR), Rubona, with cuttings from Venezuela, Uganda, Kenya, Indonesia, Zaire, etc. They provided valuable background on the factors affecting cassava adaptation, and they served as the basis for chemical analyses, yield trials, and selection.

Researchers tested several varieties to ascertain the relative concentrations of hydrocyanic acid in the roots, classifying them as:

- Bitter, very toxic, containing from 0.01872% to 0.02246% HCN, e.g., Eala Amer 07;
- Semi-sweet, moderately toxic, containing

[1] Institut des sciences agronomiques du Rwanda (ISAR), B.P. 121, Kigali, Rwanda.

from 0.01123% to 0.01591% HCN, e.g., Ikiela, Sao Pedro-Preto, Creolinha; and

- Sweet, slightly toxic, containing from 0.00561% to 0.01029% HCN, e.g., Kenya 08.

The findings indicate that certain types of cassava considered sweet in their area of origin are toxic in Rubona or moderately toxic and vice versa. For example, Amer (6 months) is bitter in Mulungu but sweet in Karama-Maza. Thus, the degree of toxicity of a variety appears to be considerably influenced by environmental conditions.

Chemical analyses have shown that cassava roots are an incomplete food, high in carbohydrates but deficient in amino acids. The leaves have not been analyzed in Rwanda because they are not yet widely used as a food.

Yield trials of cassava varieties have been carried out at Rubona since 1940. Two trials were done from 1940 to 1944, the aim being to catalogue the varieties according to fresh root, peeled root, and starch production. The variety Eala Amer 07 produced the best average yields: fresh roots, 20.4 t/ha; peeled roots, 16.7 t/ha; and starch, 4.2 t/ha. Other varieties that were also shown to be of value were: Ntolili, Sao Pedro-Preto, Creolinha, Ikiela Pacarae, Kenya 08, and Kenya 03.

Later, work on cassava at Rubona was limited to maintenance of the eight acceptable varieties in the collection and to multiplication and distribution work on the elite variety Eala 07. The intercropping of cassava with legumes, especially groundnuts, was determined to be a good practice.

Since 1975, cassava has held the attention of researchers at ISAR because of the problems hindering its production, such as: the loss of the Eala 07 multiplication fields after elephants went on the rampage at Karama in 1975; a severe attack of cassava mosaic disease; and the appearance of the green mite and bacterial blight.

The loss of the multiplication fields at Karama, which had been supplying rural farmers with improved planting material, was a considerable setback, forcing a halt to distribution while Rubona staff reestablished production. Cuttings were collected from farmers, especially those in the Icyanya and Nkanga regions where the distribution of cassava had previously been intensive, and about 20 varieties from INERA (Zaire) were added to the collection.

In November 1975, a hectare of Eala amer 07 was planted at Karama. Unfortunately, when the cuttings took root, most showed signs of mosaic disease and had to be removed from the field and burned. The remainder were monitored weekly for the presence of mosaic symptoms. By October 1976, only about 15% of the plants were left.

The cuttings taken from this material were planted (1.5 ha) in November 1976 and were screened for disease during the following year. About 50% of the plants remained healthy, providing enough planting material in November 1977 for 5 ha. Still, some of the plants (about 1%) showed symptoms of mosaic disease. These were immediately eliminated and replaced. Thus, by 1978, the ISAR station at Karama was once again able to launch its cassava-distribution program to rural areas, and by 1980 it had distributed 560 000 cuttings of Eala Amer 07.

Follow-up has shown that fewer than 10% of the plants show symptoms of mosaic disease and that it is possible to reduce this figure if farmers weed out diseased plants.

Mosaic is the main disease affecting cassava in Rwanda; its control depends on careful selection of planting material. Farmers and extension workers alike should ensure that cuttings for new plantations are from healthy stalks.

Future prospects

Cassava production in Rwanda is constrained primarily by diseases and pests such as mosaic, bacterial blight, cercosporiosis, and the cassava green mite, which has become very damaging in recent years, especially in the dry season. As a first step in dealing with these problems, ISAR has outlined a two-part, cassava-research program.

Short-term research

The purpose of the short-term research within the program is to determine the morphological and agronomic characteristics of the varieties in the collection. At present, the collection includes 55 varieties, and others will be added gradually. The results of this work are expected to be used in assisting individual farmers to choose the most suitable variety for them.

Activities began 2 years ago at Karama and Rubona and have involved the setting up of a yield-testing program for the different varieties. Harvesting is being done after 12, 18, and 24 months. At harvest, observations on incidence of disease, morphology, and agronomic characteristics will be recorded, with IITA standards being applied for shape, length, circumference, colour of peel, etc.

Long-term research

The selection process is long term and is used to obtain resistant varieties, which are vital in the search for an effective solution to disease and pest problems, and it needs to be based on a collection with wide genetic variability, produced through techniques of rapid multiplication. Thus, at present, ISAR staff are expanding their collections, three series of cassava seed beds already having been built at Karama from IITA seeds and from seeds harvested from local material. Three hundred and thirty-five cuttings are being evaluated. The purpose of the long-term research work is to obtain varieties that yield well, mature early, resist attack by diseases and pests of economic importance, and adapt readily to different cassava-producing regions.

The procedures follow a 6-year schedule. In the 1st year, seeds (60 000–100 000) are acquired from IITA and from local varieties and are planted in the Karama seed beds; sorted for resistance to mosaic, bacterial blight, cassava green mite, and other diseases of economic importance; taste-tested for level of HCN contents; and studied in terms of root structure and characteristics (shape, length, circumference, and colour of peel).

In the 2nd year, cuttings at Karama are evaluated, researchers choosing the best 2000–3000 cuttings from the best families, verifying the seed-bed evaluation for disease resistance and root characteristics, determining dry matter (expressed as a percentage), measuring yields, assessing damage from insect attack, and noting morphological and agronomic characteristics.

The 3rd year is devoted to preliminary yield trials, at Karama, of the best 100 cuttings from which about 50 are chosen for advanced yield trials the next year at three different locations (Karama, Rubona, Mayaga) and in mixed-cropping systems with beans and groundnuts.

During the 5th year, uniform yield trials are conducted with the best 25 cultivars at five different locations (Karama, Rubona, Gisaka, Mayaga, and Mutura), the results being compared with those from advanced yield tests. The last year is spent on rapid multiplication and distribution of the best 2–5 varieties.

This method of research, based on IITA standards, combines the best families with a selection of the best subjects within the best families. We at ISAR hope that, by carrying out this research with the cooperation of IITA and with other national programs, we can devise solutions to cassava-production difficulties.

Agronomic effects and economic importance of fertilizers on yams in Cameroon

S.N. Lyonga[1]

In fertilization trials in Cameroon, three elite cultivars (Batibo, Jakiri, and Oshie) of the three main species of *Dioscorea* in the country showed economic responses to nitrogen and potassium, and the best results were obtained when one application corresponded with the period of maximum plant metabolism (May–July). Generally, there was no response to phosphorus. Nitrogen (440 kg/ha) increased yields of Batibo (*D. cayenensis*) by 17.7%, Jakiri (*D. dumetorum*) by 25%, and Oshie (*D. rotundata*) by 21%. A lower level of N also gave significant yield responses. The interactions N × K and N × P were significant for Oshie but not for Jakiri or Batibo, and Jakiri generally responded to NPK interaction better than the other two cultivars. Nitrogen applied in July (the peak growing period, 60 days after planting) gave better results than that applied earlier (at planting) or later (October). Costings of fertilizer use showed profit margins even on some of the treatments that were not statistically significant. A return/cost ratio of more than 2 was regarded as profitable.

Les trois meilleurs cultivars (Batibo, Jakiri et Oshie) des trois principales espèces d'igname *Dioscorea* ont, au cours de tests d'amendement, vu leur production s'accroître avec l'azote et le potassium, les meilleurs résultats étant obtenus avec application lors de la période de métabolisme maximum (de mai à juillet). Le phosphore n'a généralement pas eu d'effet. L'azote (440 kg/ha) a accru les rendements du Batibo (*D. cayenensis*) de 17,7 %, du Jakiri (*D. dumetorum*) de 25 % et d'Oshie (*D. rotundata*) de 21 %. Une quantité inférieure de N a aussi donné des résultats significatifs. Les combinaisons N × K et N × P n'ont donné de résultats que pour l'Oshie. Le Jakiri a en général mieux réagi à la combinaison NPK que les autres cultivars. L'azote a donné de meilleurs résultats en étant appliqué en juillet (période de plus forte croissance, 60 jours suivant la plantation) que plus tôt (à la plantation) ou plus tard (en octobre). L'estimation du prix de revient des fertilisants a révélé des profits, parfois même là où les résultats n'étaient pas statistiquement significatifs. On a considéré comme profitables des revenus au moins deux fois plus élevés que les coûts.

Most peasant producers of yams (*Dioscorea* spp.) throughout the world do not use fertilizers to any appreciable extent (Coursey 1967). Perhaps the main reason is that yams are usually first in the cropping sequence in yam-producing localities and the soils have not been exhausted by other crops; other reasons are the lack of information on the use of fertilizers and the inability of farmers to afford them. The few farmers who do use organic matter in yam production primarily aim to provide mulch rather than manure. The nutrients from the rotting of the matter are secondary to the farmers, even though the value of nitrogen (Vine 1953; Doku 1967; Ferguson 1970; Umanah 1973) and, to a lesser extent, potassium (Irving 1956; Ferguson 1970) has been documented. Koli (1973), for example, reported yield increases for white yam (*D. rotundata*) of 22.1% when 67.2 kg N/ha was applied, and Enyi (1970), in his trials with Chinese yams (*D. esculenta*) in Tanzania, showed that nitrogen had positive effects on the leaf-area development and mean relative growth rate. Yam species and varieties have been found to respond differently to nitrogen fertilizers (Ferguson and Haynes 1970), but there is little information available on varietal response to potassium.

The findings on yam response to phosphorus have not been consistent. Whereas Irving

[1] Cameroon National Root Crops Improvement Program (CNRCIP), Institut de la recherche agronomique (IRA) Ekona Centre, P.M.B. 25, Nuea, South West Province, Cameroon.

(1956) and Umanah (1973) reported negative responses, Vine (1953) and Nye (1954) found no response at all. Thus, there is a need for more information about the chemistry of this element in the soil with respect to yam nutrition and yield response.

Where NPK has been judiciously used, positive results have been obtained (Mann 1963), and Wood (1933) and Stephens (1956) found a positive response when organic manure was applied to yams. This response probably reflects not only the release of nutrients by the organic matter but also other beneficial effects on the soil's physical and chemical properties such as water retention and aeration. Also, because yams are long-season crops, they need a continuous, but gradual, supply of nutrients — a prominent characteristic of the decay of organic matter.

Materials and methods

From 1971 to 1974, a series of fertilization experiments was conducted with yams at Bambui plain, which is 1330 m above sea level and is characterized by humic acrisol soil, according to USDA (United States Department of Agriculture) soil taxonomy. The objective was to document the effects on yam of N,P,K alone and in combinations.

Three elite cultivars — Batibo yellow yam (*D. cayenensis*), Jakiri trifoliate yam (*D. dumetorum*), and Oshie white yam (*D. rotundata*) — were used for the early (1971–72) trials (Table 1). The procedures for these trials were similar, although planting, fertilizing, and harvesting dates differed. The fields had previously been under grass fallow, and, after land preparation, the yam sets for each cultivar were planted on ridges (1 m × 1 m) in a 3^3 factorial design, replicated twice — a total of 54 plots for each cultivar, an area of 65 m^2.

Table 1. Fertilizer treatments (kg/ha) applied to Batibo yam (*D. cayenensis*) in 1971 and Oshie white yam (*D. rotundata*) in 1972.

	D. cayenensis	*D. rotundata*
N_0	0[a]	0[a]
N_1	220	176
N_2	440	352
P_0	0[b]	0[c]
P_1	250	220
P_2	500	440
K_0	0[d]	0[d]
K_1	200	200
K_2	400	400

[a] Urea (45% N_2).
[b] Bicalcium phosphate (38–40% P_2O_5).
[c] Single superphosphate (18% P_2O_5).
[d] Chloride of potash (60% K_2O).

The fertilizers for Batibo and Jakiri were broadcast at planting on the tops and sides of the ridges and were mixed with the topsoil by hand forks. The method of application was the same for Oshie, but only superphosphate was applied at planting. The other fertilizers were applied in equal doses 60 and 120 days later. Planting dates for Batibo and Jakiri were 11 and 29 March 1971, respectively, and Oshie was planted the next year. Harvesting was in December for Batibo and mid-October for Jakiri.

The setup for the later fertilizer trials (1973–74) was slightly different. Oshie was the only cultivar tested, and the design was a randomized, complete block, replicated six times, with an experimental plot size of 48 m^2 for the 1973 trial and 33 m^2 for that in 1974. The yam sets were planted on prepared land at 1 m × 1 m, as for the earlier trials. The superphosphate was applied at planting, and the urea and chloride of potash were split into two equal doses, applied at 60 and 120 days after planting. In the 1974 trial, which aimed to test the effects of varying the date and the amount of nitrogen, urea (220 or 440 kg/ha) was applied in 1, 2, or 3 doses (at planting, at 60, at 120 days). The amounts of P(500 kg/ha) and K(300 kg/ha) were uniform. The cost of urea was calculated at the farmers' subsidized price of 33 francs CFA/kg.

During the trials, the yams were attacked by shoe-string mosaic and anthracnose. Anthracnose was controlled by Manesan 80 (5 kg/ha sprayed in five split applications from May until the heavy rains). Shoe-string mosaic was not controlled, as there was no available information on either its epidemiology or its control.

Results

1971–72 experiments

Yields were generally good in 1971 owing to good weather conditions and little incidence of diseases and pests. This was not the case in 1972 when drought and disease seriously reduced yields. The overall differences in yield notwithstanding, the effects of fertilization were clear in both years (Table 2).

The data indicated varietal differences in response, although the variation may reflect the difference in time of applications. Batibo's response to the higher level (440 kg/ha) of nitrogen was significant at 5%, a yield increase of 17.7% over the control (no nitrogen), but its responses to the other applications were not significant. Jakiri showed significant responses to nitrogen at both levels, increases of 20% ($P < 0.05$) and 25% ($P < 0.01$) for 220 kg/ha and 440 kg/ha, respectively. There was no significant response to potash, and the higher level of phosphorus caused a depression in yield. Interactions between nitrogen and phosphorus ($N \times P$), nitrogen and potash ($N \times K$), and phosphorus and potash ($P \times K$) were not significant for Jakiri, but the high order interaction NPK was significant, representing a yield increase of 26.6% by $N_1P_2K_2$ and 15% by $N_2P_2K_1$ over the control. Oshie responded significantly to both levels of nitrogen and potash, $P < 0.05$ for the lower levels and $P < 0.01$ for the higher amounts. The yield increases over the control were 11.1% and 22.1% for nitrogen at 220 and 440 kg/ha and 13.4% and 22.6% for potash at 176 and 352 kg/ha. The interactions between nitrogen and phosphorus ($N \times P$) and nitrogen and potash ($N \times K$) were significant at the 5% level. There was no response to phosphorus alone, and neither the interaction between phosphorus and potassium nor that with NPK was significant.

Economic considerations

The costs of each fertilizer input were calculated as background for the economic perspective of the variance of yields (Table 3); they were based on the cultural practices used (3 workdays/ha for each application at 200 francs CFA/workday; transportation and charges of 50 francs/50-kg bag of fertilizer from the store to the field; and 15 francs/kg of fresh white and yellow yam tubers and 10 francs for trifoliate yam). These calculations indicated clearly the economic benefits of a sample of fertilizer treatments, including some of those that were not considered to be significant in the analysis, for example, potassium (220 kg/ha) on *D. dumetorum*. (When the ratio of extra value to extra cost — that is, extra returns/fertilizer inputs — was greater than 2, the treatment was considered profitable.) This fact underlines the im-

Table 2. Effect of fertilizer treatments on yield (t/ha) of (A) yellow yam (*D. cayenensis*), (B) Jakiri trifoliate yam (*D. dumetorum*), and (C) Oshie white yam (*D. rotundata*).

Treatment	Yield		
	(A)[a]	(B)[b]	(C)[c]
N_0	19.8	23.2	16.6
N_1	22.1	17.9[d]	18.5[d]
N_2	23.3[d]	29.0[e]	20.1[e]
N_0P_0	20.8	25.8	16.2
N_0P_1	19.5	23.6	18.8
N_0P_2	19.8	20.2	15.0
N_1P_0	19.6	26.0	19.3
N_1P_1	22.0	30.0	19.0
N_1P_2	24.6	27.5	17.0
N_2P_0	21.3	26.7	18.7
N_2P_1	23.7	32.2	19.9[d]
N_2P_2	25.1	28.2	21.0[d]
$N_0P_0K_0$	19.2	28.9	14.8
$N_1P_1K_1$	26.4	33.1	18.2
$N_1P_1K_2$	19.0	27.7	22.1
P_0	20.5	26.2	18.0
P_1	21.4	28.6	19.3
P_2	23.2	25.3	16.6
N_0K_0	20.7	23.2	16.3
N_0K_1	19.4	24.7	15.6
N_0K_2	19.3	21.7	18.0
N_1K_0	23.4	26.8	15.9
N_1K_1	21.9	28.0	18.0
N_1K_2	20.6	28.9	21.5[d]
N_2K_0	22.5	25.2	17.2
N_2K_1	22.5	31.3	22.1[d]
N_2K_2	25.0	30.6	21.0[d]
$N_1P_2K_1$	21.5	25.5	16.8
$N_1P_2K_2$	24.8	36.6[e]	21.4
$N_2P_1K_1$	25.3	31.8	20.3
K_0	22.2	25.1	16.4
K_1	20.4	27.9	18.6
K_2	21.7	27.1	20.1[e]
P_0K_0	20.3	26.6	15.7
P_0K_1	19.6	26.2	18.3
P_0K_2	21.7	25.7	20.3
P_1K_0	21.3	27.4	18.0
P_1K_1	22.8	30.5	18.5
P_1K_2	20.1	27.8	21.3
P_2K_0	25.0	21.2	15.7
P_2K_1	21.2	27.1	18.9
P_2K_2	23.4	27.6	18.8
$N_2P_1K_2$	22.6	32.7	20.2
$N_2P_2K_1$	23.8	33.3[d]	26.2
$N_2P_2K_2$	26.6	27.5	20.4

[a] Standard error = ± 5.8 kg; coefficient of variability = 15.68%.
[b] Standard error = ± 7.3 kg; coefficient of variability = 16.075%.
[c] Standard error = ± 3.54 kg; coefficient of variability = 11.32%.
[d] Significant at 5% level.
[e] Significant at 1% level.

Table 3. Economic returns of application of mixed fertilizer to three yam species.[a]

Treatments	Return/ha (1000 francs CFA)	Return/extra cost
D. cayenensis		
N_1	34.5	5.6
N_2*	52.5	4.5
N_2K_2	64.5	2.9
$N_1P_1K_1$	108.0	5.7
D. dumetorum		
N_1	47.0	7.6
N_2*	58.0	5.0
K_1	28.0	4.8
$N_1P_1K_1$	42.0	2.2
$N_1P_2K_2$**	77.0	2.4
D. rotundata		
N_1*	28.5	5.1
N_2**	52.5	5.3
K_1*	33.0	5.1
K_2**	55.5	4.8
N_1P_1*	42.0	3.6
N_2P_2*	72.0	3.3
N_1K_1	25.5	2.4
N_1K_2*	78.0	4.9
P_1K_1	42.0	3.3
P_2K_1	48.0	2.7
$N_1P_1K_1$	51.0	3.0
$N_2P_2K_2$	84.0	2.6

[a] ** = significance in yield at 1%; * = significance in yield at 5% level.

portance of economic considerations in such trials because they are what actually decide whether or not research results will be applied.

1973–74 experiments

As in 1972, the dry season in 1973 started suddenly (6 weeks earlier than usual, in early October). The situation must have forced maturity of the yams, thus reducing the period of tuber bulking and, consequently, yields. Anthracnose, too, probably affected the yields, despite control measures.

Of the six NPK treatments on Oshie, five were associated with yields that were significantly ($P < 0.05$) larger than those of the control (Table 4). Yield increases ranged from 19% to 31.9%; the response to the fertilizers was obtained from the potash and nitrogen components.

The extra value provided by the yield increases was calculated on the basis of 1973 prices (white yam had increased from 15 francs to 22, whereas labour and transportation costs had remained constant), and the results were compared with the costs of the inputs. Four treatments (Table 4) had an extra value/cost ratio greater than 2 and were, thus, definitely profitable. One treatment ($N_2P_1K_2$), with a value of 1.7, was marginal, and one treatment ($N_2P_1K_1$), with an increase of only 1 t/ha over the control had an extra value/cost ratio less than 1, which translated into a loss of 1500 francs CFA.

In the experiments to test the effects of different dates (at planting/60 days/120 days) and levels (kg/ha) of nitrogen application, three treatments (0/440/0, 110/110/0, and 0/110/110), all of which included some fertilizer in July, the period of rapid leaf-area development and tuber initiation, yielded significantly higher than the control, with increases of 26.7%, 20.7%, and 19.3% (Table 5). These results supported some of the findings of Chapman (1965) with *D. alata*, and Koli (1973), Umanah (1973). In fact, all seven treatments that produced yields higher than the control included fertilizer applications in July, whereas the four that produced yields below the control had applications either in April or in October.

The economic situation was, however, different. Five treatments (0/440/0, 110/110/0,

Table 4. Yield and economic response of *D. rotundata* to mixed NPK fertilizers (as single treatments).

Treatment	Yield (t/ha)[a]	Return/ha (1000 francs CFA)	Return/ extra cost
$N_1P_2K_2$	15.3a	81.4	2.7
$N_2P_2K_2$	15.2a	79.2	2.2
$N_1P_1K_2$	14.7a	68.2	2.8
$N_2P_2K_1$	14.5ab	63.8	2.1
$N_2P_1K_2$	13.8b	48.4	1.7
$N_2P_1K_1$	12.6bc	22.0	0.9
$N_0P_0K_0$	11.6c	0	0

[a] Numbers followed by a different letter are significantly different at 5% level (Duncan's test).

Table 5. Effect of split application of urea on yield and economic returns from Oshie white yam (*D. rotundata*).

Treatment[a]	Yield (t/ha)	Return/ha (1000 francs CFA)	Return/ cost
0/440/0	19.0[b]	100.0	6.1
110/110/0	18.1[b]	77.5	8.4
0/110/110	17.9[b]	72.5	7.9
0/220/0	17.5	62.5	7.3
110/220/110	17.1	52.5	3.0
220/220/0	15.9	22.5	1.3
0/220/220	15.3	7.5	0.4
0/0/0 (control)	15.0	0	0
220/0/0	14.4	-15.0	–
0/0/440	14.4	-15.0	–
440/0/0	14.0	-25.0	–
0/0/220	13.2	-45.0	–

[a] Indicates amount of urea applied (kg/ha) in April/July/October.
[b] Significantly different (5%) from control.

0/110/110, 0/220/0, 110/220/110) appeared profitable. One treatment (220/220/0), with an extra return/extra cost ratio of 1.3, was marginal, and the other five treatments were uneconomical.

Discussion and conclusions

Yellow yam's low response (Table 2) to fertilization may be attributed partially to the fact that all the fertilizers were applied at once, at planting time. Much of the soluble nitrogen and potassium may have been lost through leaching and runoff before the plants were ready to make use of it. The soil contained relatively low potash (0.37 meq/100) at the surface horizon, and one would have expected a response to this nutrient.

Nevertheless, all three elite cultivars of the three species showed some response to nitrogen in the humic acrisols of Bambui plain. This result agrees with the findings of Obi (1959) in Nigeria and Doku (1967) in Ghana.

The 22.6% increase in yield of Oshie white yam in response to potash (400 kg/ha) can partly be attributed to the split applications (May and the end of June). The interactions of N × P and N × K were positive. There was no response to phosphorus alone in any of the trials.

Fertilizer use, especially low levels of nitrogen, potash, and NPK, appeared to be very economical, giving an extra return/extra cost ratio of between 2.2 and 8.4 (including some of the treatments that were not statistically better than the control). This finding suggests that experimental statistical analysis of data is not always consistent with economics and needs to be supplemented by cost analysis.

The data from this series of studies suggest that, in the humic acrisols of the grassland highlands of Cameroon, fertilization of *D. rotundata* with nitrogen (total amount: 220–440 kg/ha), applied at planting (April) and during the period (July) of maximum leaf-area development and tuber initiation, produces the best results. This approach may also prove effective for *D. cayenensis* and *D. dumetorum*.

The results with split applications of fertilizer on Oshie white yam indicate the value of linking fertilizer work with growth studies of promising cultivars of different species so that factors responsible for greater yields can be coordinated most effectively.

It should be noted, however, that the recent rise in the cost of fertilizers is a serious limitation to their use. Unfortunately, it comes at a time of greater awareness among farmers about the importance of fertilizer use. The move by some African governments to try to reduce costs by producing fertilizers locally or by subsidizing the price is welcome.

Country reports

Colocasia *in a market in Cameroon — a reminder that research must produce results acceptable to consumers.*

Cameroon

H.J. Pfeiffer and S.N. Lyonga[1]

This paper describes the cropping systems found in the different zones in Cameroon and relates them to the ecological and sociological conditions of the regions. A large variety of soils and ecological zones characterize the country, and the potential for root-crop production ranges from poor to excellent.

Description des systèmes agricoles des différentes régions du Cameroun et de leurs liens avec les conditions écologiques et sociales. Le pays est caractérisé par une grande diversité de sols et de zones écologiques dont le potentiel pour la production de tubercules alimentaires va de faible à excellent.

The most important root and tuber crops, except for some yams, were introduced quite recently into West and Central Africa. However, they are now fully integrated into the traditional farming systems and constitute a staple food for the population of Cameroon, their production outstripping that for cereals, legumes, and other crops (Table 1). Nearly all production comes from smallholders, and, because of the difficulties in ascertaining the outputs, production figures must be regarded as approximations. The traditional system is mixed cropping; therefore, yields, such as those calculated by Letouzy in the 1950s, are based on low plant density and would have to be corrected for evaluations of potential yields. The soils in Cameroon are varied and include both poor and good potential for root-crop production (Fig. 1, Table 2).

Lake Chad basin (zones 1–5, 7, 8)

The principal factors limiting root and tuber production in the Lake Chad basin are bad drainage, coupled with periodic flooding (zones 1–4), soil alkalinity (zones 1, 4, 5), low rainfall (<900 mm), and the limited number of days of rain.

Cassava and sweet potato, generally in pure stands, can be found on the hydromorphic soils in the basin (zone 3) after the heaviest rains and on soils characterized by good water retention (zone 7). Yams are only found on very fertile soils with a high water-retention capacity because of the high nutrient requirements of this crop.

Ferruginous (zones 6, 9–13, 14a, 15)

The ferruginous zone is characterized by good drainage, except in the swamps, and

Table 1. Production and cultivated area of various crops in Cameroon 1978–79.[a]

	Production (10^3 t)	Cultivated area (10^3 ha)
Cereals	408.9	490.4
Legumes	103.9	871.7
Root and tuber crops	1958.2	1679.5
Cassava	632.2	545.4
Cocoyam + taro	815.4	580.1
Yam	417.3	326.2
Sweet potato	68.0	159.2
Irish potato	25.3	68.6
Bananas	3103.2	1232.1
Other crops	33.1	76.5

[a] Extracted and adapted from Cameroon, Ministère de l'agriculture, 1979. Yearbook of agricultural statistics 1978–1979. Yaoundé, Cameroon, Ministère de l'agriculture.

[1] Cameroon National Root Crops Improvement Program (CNRCIP), Institut de la recherche agronomique (IRA), B.P. 2067, Yaounde, and IRA Ekona Centre, P.M.B. 25, Nuea, South West Province, Cameroon.

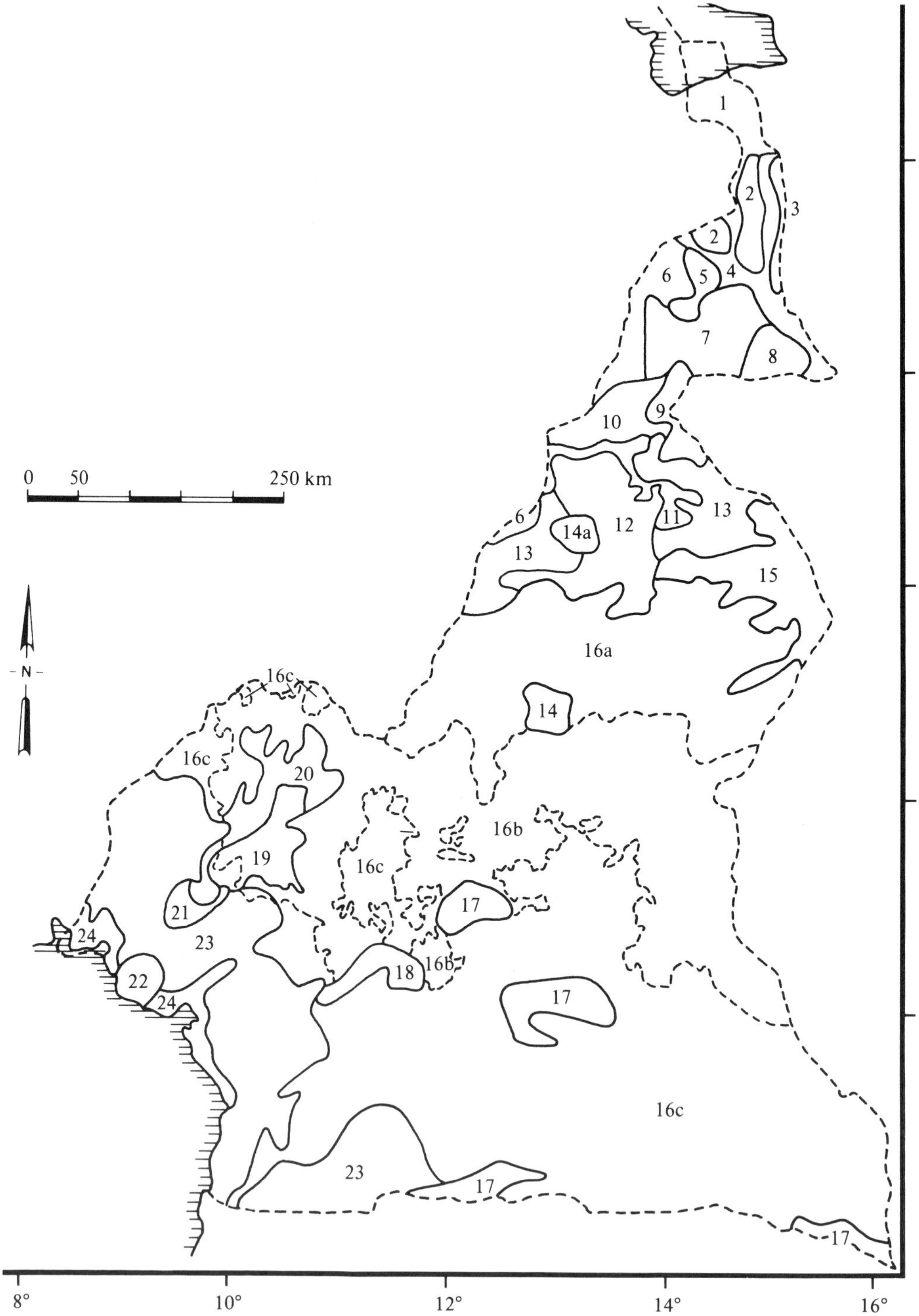

Fig. 1. Soil zones found in Cameroon.

medium to low fertility. Rainfall varies between 800 mm and 1500 mm, and the number of days of rain can reach 125. Therefore, this Sudan savanna is better suited to root- and tuber-crop production than is the sahelian zone of the basin.

Monocropping in large plots (≥1ha) is common for short-cycle (6–8 months) cassava. All of the root and tuber crops (cassava, macabo, taro, yam, sweet potato) are found inside the house compound and are cultivated with cereals and vegetables. In swampy land, sweet potatoes are sometimes cultivated on ridges at the end of the rainy season.

One of the most important yam-production centres in Cameroon is the Mbe plain (south of zone 12). Yam is grown in pure stands on the well-drained sandy soils, which are easy to mound. Finally, Irish potatoes are cultivated with success in the higher elevation areas found in zone 6.

Ferralitic (zones 16–20, 23)

The ferralitic zone soil is characterized by low to medium fertility, good drainage, and favourable physical conditions in the absence of a ferruginous hard pan at low depth. The exceptions to this generalization are the andosols, which are fertile, and some hydromorphic areas, gleysols, and histosols where root and tuber crops are rarely grown. At the phytogeographic level, this zone should be divided into five areas.

Adamaoua Sudano-Guinea bush savanna (zone 16a)

The Adamaoua Sudano-Guinea is a ranching area and is more favourable for cassava growing than is the ferruginous area because of its more humid climate. Cassava is planted mostly in pure stands on large areas (>1 ha). Because rainfall is sufficient, the crop can be maintained for 15–18 months (sometimes 2 years) on the land. Planting is normally done at the beginning of the rainy season; however, very young plantations can also be found at the beginning of the dry season. Sweet potatoes are cultivated in small fields on ridges in pure stands.

In most cases, the other root and tuber crops are cultivated around the houses and mixed with some cereals and local vegetables. This "garden" is protected from animals by a fence.

The importance of cassava compared with cereals increases toward the south: the southeast and southwest parts of zone 16a are among the largest cassava-producing areas on a per-person basis.

Postforest Congo-Guinea savanna (zone 16b)

Cassava is quite important in the transition area between the forest and the savanna, especially in the eastern and northern parts. In the savanna, cassava is monocropped or relay-cropped with peanuts or maize. Cocoyam and yam are mostly cultivated on newly cleared forest (galleries) because they have greater nutritional requirements than cassava does. In the northeastern savanna, the "ecobuage" technique is commonly practiced. Taro and cocoyam are intercropped with peanuts, maize, and melon. The principal yam-production centres are in the south (Bafia) and in the northeastern part of this zone. Cocoyam cultivation is more common in the southern part.

Semideciduous forest (zones 16c, 17, 18)

Mixed cropping and relay cropping are the usual cultivation practices in the semideciduous forest. The area cultivated by a single farmer is usually relatively small (1000–5000 m^2) because of the difficulty in clearing the land. After the trees are felled and burned, peanuts are sown and allowed to become established. A superficial hoe plowing is done, and, under this cover, maize is sown in pockets (3–5 stands) at a distance of 1–2 m in each direction. Finally, cassava stakes are planted symmetrically in relation to the already sprouted maize. Some plantain, yam, and cocoyam will also be inserted. On fertile land, melon and other vegetables are integrated as well.

After the 1st season, peanuts, maize, melon, and legumes are harvested. Later, when the cassava has been harvested, the field is either left fallow or planted again with cassava or sweet potatoes. The harvest is done according to the needs of the family, and almost no weeding is done; thus, most of the cassava harvested during the 2nd year is in a eupatorium field.

Table 2. Soil type, rainfall, altitude, soil characteristics, and root and tuber production in different regions of Cameroon.

Zone	Soil type	Rainfall mm	Rainfall days	Altitude (m)	Soil characteristics: Drainage	Soil characteristics: Fertility	Root and tuber production (10^3 t): Cassava	Macabo/ taro	Yam	Sweet potato	Potato
Lake Chad basin zone											
1	Planosol (solodic)	<600	<75	300	Rather bad	Alkaline (Na)	–	–	–	–	–
	Vertisol (pellic)				Bad	Very high	–	–	–	–	–
	Fluvisol (eutric)				Rather bad	Medium	little	–	–	little	–
2	Luvisol (gleyic)	600–750	<75	300	Very bad	High	–	–	–	–	–
3	Gleysol (eutric)	600–750	<75	300	Bad	Medium	4.1	0	0	0.16	0
4	Vertisol (pellic)	650–850	<75	300–600	Rather bad	Alkaline (Na)	–	–	–	–	–
	Gleysol (eutric)				Very bad	Medium	–	–	–	–	–
5	Planosol (solodic)	700–800	<75	300–600	Bad	Alkaline (Na)	–	–	–	–	–
7	Vertisol (pellic)	750–900	<75	300–600	Bad	Very high	–	–	–	–	–
	Vertisol (chromic)				Fairly good	High	2.4	0	0.15	1.7	0
	Gleysol (eutric)				Rather bad	Medium	–	–	–	–	–
8	Arenosol (chromic)	800–900	75	300–600	Excessive	Low	0.25	0	0	0.6	–
	Gleysol (eutric)				Rather bad	Low	–	–	–	–	–
Ferruginous zone											
6	Regosol (eutric)	800–950	75	600–1000	Good	Low	3.1	0.9	1.3	1.8	1.9
	Luvisol (ferric)				Good	Medium	19.5	0.5	4.8	2.6	0.4
9	Luvisol (chromic)	850–1000	75	600–1000	Good	High	–[a]	–[a]	–[a]	–[a]	–[a]
10	Planosol	900–1000	75	600	Good	Low	–[a]	–[a]	–[a]	–[a]	–[a]
	Luvisol (ferric)				Good	Medium	–[a]	–[a]	–[a]	–[a]	–[a]
	Vertisol (chromic)				Fairly good	High	–[a]	–[a]	–[a]	–[a]	–[a]
11	Fluvisol	1000–1250	75–100	300	Bad	High	–[a]	–[a]	–[a]	–[a]	–[a]
12	Luvisol (ferric)	1000–1500	75–125	600	Good	Low	–[a]	–[a]	–[a]	–[a]	–[a]
13	Luvisol (gleyic)	1000–1250	75–100	600	Rather bad	Low	–[a]	–[a]	–[a]	–[a]	–[a]
	Luvisol (ferric)				Good	Low	–[a]	–[a]	–[a]	–[a]	–[a]
14a	Lithosol (ferric)	1250	100	600–1500	Good	Low	–	–	–	–	–
15	Luvisol (plinthic)	1250–1500	100–125	300–900	Good	Medium	–	–	–	–	–

Ferralitic zone											
14	Lithosol	1250	100	600–1500	Good	Low	–	–	–	–	–
	Ferrasol (orthic and rhodic				Good	Medium-low					
	Nitosol and aerosol										
	a. Sudano-Guinea Savanna (Adamaoua)	1250–1600	100–150	900–1500			124.0	0.89	1.89	2.0	little
	b. Postforest savanna (galleries)	1500–1750	125–150	600–900			78.3	26.4	44.78	12.9	0.6
	c. Semideciduous forest	1500–1750	125–175	400–900			148.1	133.2	18.53	7.5	3.8
	Gleysol (dystric)	1500–1750	125–175	400–900	Very bad	Low	–	–	–	–	–
17	Ferralsol (orthic)	1500–1750	125	600–900	Good	Medium-low	–[b]	–[b]	–[b]	–[b]	–[b]
	Histosol (dystric)				Very bad	Low	–[b]	–[b]	–[b]	–[b]	–[b]
18	Gleysol (humic)	1500–1750	125–175	300–600	Bad	Low	23.5[c]	11.4[c]	8.9[c]	0.3[c]	little
	Nitosol (dystric)				Good	Low					
19	Nitosol (dystric)	1750–2500	150–200	600–2000	Good	Low					
	Andosol (humic)				Good	Excellent	23.1[c]	220.0[c]	157.7[c]	19.9[c]	8.9[c]
	Gleysol (humic)				Bad	Low					
20	Ferralsol (humic)	2000	150–200	600–2000	Good	Low	74.4[c]	161.8[c]	90.2[c]	16.3[c]	9.3[c]
	Andosol (vitric)				Good	Good					
23	Ferralsol (xanthic)	1750–4000	175–225	400	Good	Low	76.1	170.2	67.8	1.4	0.65
Volcanic zone											
21	Cambisol (humic)	2500–4000	200	100–2000	Good	Good	38.8	70.3	16.2	3.1	little
22	Andosol (humic)	2500–4000	200	600–2000	Good-medium	Good	12.7	19.8	4.7	0.1	–
Coastal alluvial zone											
24	Gleysol (dystric)	3000	225	100	Mangrove soils		–	–	–	–	–

[a] Yields are included in 6 — Luvisol (ferric).
[b] Yields are included in 16c.
[c] Yields are totals for area (18, 19, and 20 respectively).

Variations in this general cropping pattern exist and mostly reflect local food habits. For example, in the Bassa area, cocoyam is widely preferred to cassava. Sweet potatoes are normally only grown among vegetables, yam, and plantain in small areas around the houses, except in zone 18 where pure stands (200–300 m^2) are cultivated on mounds. Cocoyam is also cultivated as a garden crop where it can benefit from organic household refuse. The crop is harvested according to family needs.

Mountain (zones 19, 20)

The soils in the mountainous zone have been influenced by volcanic material (zones 19a, b). Physically, they are mostly favourable, with good drainage, but, chemically, they vary. The best soils (zone 19b) are occupied by industrial coffee plantations or intensive vegetable cultivation — the food crops being grown mostly on the less fertile nitosols and ferralsols. Hydromorphic soils (zone 19c) can also be used for food crops if properly managed.

Population density and altitude determine the cropping system that is used in this area of semipermanent land occupation. Cassava is mostly planted around or in coffee fields and around houses where it is associated with maize, cocoyam, taro, beans, bananas, and vegetables. It is rarely cultivated above 1200 m, elevations where cocoyam, taro, sweet potatoes, and yams (principally *D. dumetorum*) are given priority. In very densely populated areas, cassava cultivation is reduced.

Rainfall is plentiful, and measures, such as ridging and mounding, are practiced to improve drainage. Irish potatoes gain in importance with altitude and may be considered a local cash crop.

Evergreen forest (zone 23)

Heavily desaturated, ferralitic soils that are good physically (sometimes very sandy) but poor chemically characterize the evergreen forest. The cropping system is similar to that found in zone 16c, but cocoyam and taro are important. Yam, intensively cultivated and relay-cropped with cassava, is commonly found in the sandy Muyuka plain (north of zone 22).

Volcanic (zones 21, 22)

The recent (zone 21) or older (zone 22) soils derived from volcanic material have medium physical and favourable chemical properties and are mostly occupied by industrial cash-crop plantations (bananas, pineapple, rubber, palms). Food crops are pushed back to the periphery of this zone to the temporary plantation fallows (especially for bananas).

The volcanic zone can be divided into low and high altitudes. The cropping system in the low areas is similar to that in zone 23, although the plantation fallow area is cultivated with pure stands of cocoyam (taro) or local vegetables. The importance of cocoyam and taro increases on the rich soils, whereas cassava is more important on the poorer soils. At higher altitudes, the cropping system resembles that for zone 19a.

Coastal alluvial (zone 24)

Mangrove soils are found in the coastal alluvial zone and are unsuitable for agriculture. They have only been included in this description because they are found within the country.

Kenya[1]

G.H. de Bruijn[2] *and E.J. Guthrie*[3]

Cassava is grown on about 7.0×10^4 ha in Kenya and is mainly used for human consumption, although interest in its use in animal feed and industry is growing. In some areas, it is a staple food, in others a famine reserve. Main production areas are in the west of the country and at the coast. Because of the rapid increase in population, the area of production is likely to expand, mainly in semi-arid areas. The main constraint is cassava mosaic disease (CMD), which causes serious yield reductions. The disease can be controlled in most cases by the use of healthy stakes and subsequent roguing of infected plants. Other constraints are bacterial blight and cassava green mite, both of which entered the country in recent years. These constraints should be the focus of any program to improve production. Efforts should be devoted to varietal development through selection, agronomic investigations, and sanitary measures for disease control.

La culture du manioc s'étend sur quelque $7{,}0 \times 10^4$ ha au Kenya. Il sert surtout à la consommation humaine mais on s'intéresse de plus en plus à ses applications dans l'industrie et dans l'alimentation animale. Selon les régions, il sert d'aliment de base ou de réserve en cas de disette. La production se fait surtout sur la côte et dans l'ouest du pays. Étant donné la croissance démographique rapide, la zone de culture va probablement s'étendre, surtout dans les régions semi-arides. Le principal facteur limitant de la production est la mosaïque qui provoque d'importantes baisses de rendement. Dans la plupart des cas, on peut lutter contre la maladie en utilisant des tuteurs sains et en éliminant ultérieurement les plants infectés. Mais la brûlure bactérienne et la teigne apparues au cours des dernières années causent d'importants dégâts aux cultures. Aussi, les programmes d'amélioration de la production devront-ils s'attacher en priorité à lutter contre ces ennemis du manioc, en orientant les efforts sur la création de variétés, par la sélection, des investigations agronomiques et l'application de mesures de contrôle.

Cassava occupies a relatively unimportant position in Kenya, except at the coast and in the western part of the country, where it is a staple food. According to FAO (1976), the total area under cassava in Kenya is 9.5×10^4 ha. The government (Kenya, Central Bureau of Statistics 1978) placed the total area at 7.0×10^4 ha, of which 5×10^3 ha are at the coast and 6.2×10^4 ha in the west.

Cassava occupies a smaller area than do Irish potatoes, which cover 2.6×10^5 ha, whereas sweet potatoes are reported to be grown on 3.2×10^4 ha (Kenya, Central Bureau of Statistics 1978). There is considerable potential for the expansion of cassava cultivation.

Little cassava is grown at more than 1500 m above sea level, although small amounts are found at higher altitudes, including the Nairobi area (1650 m).

A limitation on expansion is the lack of demand for cassava, which is often regarded as a famine reserve only. However, because population growth in Kenya is among the highest in the world (about 4% annually), the demand for cassava is likely to increase significantly, especially in the semi-arid areas where the yields of popular crops such as maize and field beans are unreliable. Expansion of drought-tolerant crops such as cassava

[1] This paper deals with cassava not root crops in general, even though, for example, Irish potatoes occupy greater land area. The focus is on cassava because of its importance to this meeting and its potential importance to Kenya in the future.

[2] Department of Crop Science, University of Nairobi, P.O. Box 30197, Nairobi, Kenya. Permanent address: Department of Tropical Crop Science, Agricultural University, Wageningen, The Netherlands.

[3] Overseas Development Administration, Crop Virology Research Project, Kenya Agricultural Research Institute, P.O. Box 30148, Nairobi, Kenya.

is increasingly important in such areas. Moreover, the national development plan for 1979–83 (Kenya, Government of 1979) states that, contrary to popular opinion, the nutritional factor in short supply is not protein but energy. This statement portends increased production of cassava for human food, and interest in cassava production for animal feeds and for industrial uses is also increasing.

Varieties

There are many local cassava varieties in Kenya, and a few exotic species, such as F100 and F279 from Java and Aipin Valenca from Brazil, have been grown on a limited scale. A major breeding program was initiated by Storey (1936), with the primary objective of increasing resistance to cassava mosaic disease (CMD), which causes major yield reductions in the country. He and his co-workers made crosses between *Manihot esculenta* and a number of other species, including *M. catingae*, *M. dichotoma*, *M. glaziovii*, *M. melanobasis*, and *M. saxicola* (Doughty 1958). Selections from this program are still maintained in Kenya, and they provided a major source of CMD resistance for the cassava-breeding program at the International Institute of Tropical Agriculture (IITA). A few of the best selections were also released to farmers, but the overall impact on cassava production in Kenya has been small.

Little accurate information is available on yields of local varieties in Kenya. Records of trials often include a reference to the presence of CMD and are, therefore, of doubtful value. Recent trials with mosaic-free local varieties and selected lines from the Storey program have shown that there is good potential for high yields within the local varieties (E.J. Guthrie, unpublished; De Bruijn 1980). There is little doubt that expanding efforts to develop improved varieties from local and mosaic-resistant (notably from IITA) cultivars would be productive.

Cropping systems

About 60% of the cassava is grown in pure stands, and the remainder intercropped (Kenya, Central Bureau of Statistics 1978). At the coast, and also in dry areas, cassava is mostly grown in mixtures, in most cases with maize but also with grain legumes and other crops. At the coast, cassava is also grown in association with coconuts.

Throughout Kenya, maize and field beans or pigeon peas are popular in intercropping systems. As in the semi-arid areas, in dry years, the yields of maize and beans can be extremely low and need to be supplemented by a drought-tolerant crop. Thus, the incorporation of cassava in these mixtures may increase stability of food production and is currently being investigated.

In areas where bush fallow is practiced, cassava — because of its low nutrient requirements — is often grown as the last crop before the nutrient-depleted soil is left fallow (Acland 1975). In fact, experiments at the Kenya coast have shown that cassava in a weedy fallow has a positive effect on restoration of soil fertility (Clarke 1962). These findings deserve to be followed up with studies on the optimum use of cassava in cropping patterns and rotational sequences, especially in semi-arid areas.

Time of planting and harvesting

The lack of information extends to optimum times of planting and harvesting. Time between planting and harvesting (growing season) varies widely as a reflection not only of environmental conditions but also of the purpose for which the crop is grown. For example, in areas where cassava is grown primarily as a famine reserve, plants are sometimes left in the ground for several years.

Precise information on the relationships between planting time, length of growing season, and yield is very desirable for all areas of production. Also, there is considerable scope for experimentation to determine which of the two rainy seasons is more favourable for planting and the optimum length of growing period, depending on locality and variety.

Pests and diseases

CMD is a serious problem throughout Kenya. Many plantings are 100% infected; the average level of infection in a survey of a major cassava-growing area in western Kenya was 85% (Bock and Guthrie 1978).

Estimates of yield losses from CMD infection vary widely, but it is generally agreed that losses are higher in plants derived from infected planting material than in those infected during growth. In a trial at the Kenya coast using infected cuttings, yield losses in two varieties were 70% and 86% (Bock and Guthrie 1978). The reduction in yield on a national basis may be as high as 50%; the importance of the disease and the desirability of control measures need no further stress.

In studies of the epidemiology of CMD in Kenya, Bock and Guthrie (1978) found that the rate of spread of the disease was, in general, low. The whitefly (*Bemisia tabaci*) proved to be an inefficient vector under Kenyan conditions, and human beings (by planting infected stakes) are the principal vector. The peasant farmer is generally unaware of the significance of CMD and makes no effort to select healthy planting material.

It should be possible to maintain effective control of CMD in Kenya by the use of healthy stakes during planting and subsequent roguing of plants showing symptoms of the disease, especially in the early stages of growth. A program of multiplication of CMD-free material has already been mounted, and, in the coastal areas, more than 20 ha were being grown in 1980–81. This program is likely to be expanded. The varieties under multiplication have shown some variation in degree of resistance to CMD, but the overall level of resistance appears adequate. In contrast, a collection of varieties imported from Brazil and, therefore, unselected for resistance to CMD, reached levels of infection of about 50% in 12 months when grown at the Kenya coast, an indication that the infection pressure is by no means negligible.

Two exceptions to the general effectiveness of sanitation methods have, however, been noted. In one trial at the coast, which included a core of infected plants, the rate of spread was 54% in 14 months. In a recent trial at Kampi ya Mawe in Eastern Province, levels of CMD also built up rapidly and reached 45, 54, 55, and 86%, respectively, in four varieties after 7 months. The trial contained a high proportion of a variety that is apparently relatively susceptible to CMD. However, trials at other sites using identical planting material have shown only low levels of CMD.

In addition to CMD, one other virus disease is known to occur on cassava in Kenya. This is cassava brown streak (CBSV), which is apparently confined to coastal areas. CBSV is currently of little significance; it seems that most varieties have adequate resistance. Symptoms include yellow vein banding of the older leaves and browning of the vascular tissue of the roots.

Two other major problems have recently been recorded in Kenya, namely bacterial blight (*Xanthomonas manihotis* and *X. cassavae*) and cassava green mite (*Mononychellus tanajoa*). Presence of bacterial blight was confirmed in 1978, but it had probably been present for several years (Onyango and Ramos 1978). It has not been recorded outside Western and Nyanza provinces. There are indications of differences in varietal susceptibility, but no detailed screening has yet been attempted.

Cassava green mite was first recorded in Uganda in 1971 (Nyiira 1977) and spread rapidly from the initial focus, reaching western Kenya about 1977. More recently, it has appeared on the Kenya coast, probably by spread from Tanzania, and has reached an area some 150 km inland. There are no records from Eastern Province, but these will doubtless come. Shukla (mimeographed report) found differences in varietal susceptibility to mite damage in material collected in Tanzania. Similarly, there are indications that Kenyan varieties differ markedly in degree of mite attack.

Scale (*Aonidomytilus albus*) seems to occur mainly in drier areas and is not a major problem, although it can be very damaging locally.

Utilization

Increasing population pressure and associated food shortages are likely to stimulate increased demand for cassava in the future. Consumption is likely to increase in areas where it is uncommon at present, especially in the semi-arid regions but probably also in the urban centres. Use of cassava may also be spread more evenly through the year, rather than being concentrated in periods of drought. There is scope for nutrition and home economics workers to promote more attractive and varied methods of preparation, based on local and international experience.

At present, Kenya has a shortfall in supplies of wheat. If it were possible to substitute a cassava flour for some of the wheat flour in bread making, there would be savings in foreign exchange.

Interest is developing in processing of dried cassava for use in animal feeds. Although, at present, the emphasis is for export, use in animal feed within Kenya is probably at least equally important. Selection of varieties suitable for the harvesting of foliage as high-protein fodder would be valuable, especially as a supplement for dairy cattle in the drier areas. In view of the present population trends, it seems likely that the demand for food within Kenya will ultimately override any potential benefit from the export of a relatively low-value crop such as cassava.

Factories for the manufacture of human foods, textiles, and paper are all established in Kenya, and their use of cassava starch seems a logical development. In fact, a small-scale starch factory is operating at the coast, but, at present, supplies of raw material are insufficient.

In common with most other countries, Kenya suffers from an increasing economic burden in importing supplies of petroleum fuel. Consequently, there is now a very active interest in the production of ethanol, from crops such as sugarcane or cassava, as a fuel additive. The prospects for cassava are good, as it grows in less favourable conditions than does sugarcane.

Research needs

The work on cassava in recent years has indicated the crop's potential but has also shown the need for an expanded research program that caters to the conditions and needs of all the major cassava-growing areas. Information is needed on many aspects of agronomy and farming systems, particularly on intercropping and optimum time of planting and harvesting.

An extensive program is needed on varietal development. The steps would be to make a comprehensive collection of local varieties, ensure they are free from disease (particularly CMD), and subject them to yield trials and laboratory analyses, as well as screening for resistance to pests and diseases. Some introductions from other countries would also be justified. Such a program would be followed by bulking of selected lines for release to farmers. For maximum impact, it would be an advantage to link this with a package of improved cultural practices. Above all, the importance of sanitation in the control of CMD should be stressed.

We wish to thank the Kenya Ministry of Agriculture, the University of Nairobi, and the Director, Agricultural Research Department, Kenya Agricultural Research Institute for support and permission to publish this paper.

Uganda

Z.M. Nyiira[1]

Of the major root crops, cassava and sweet potatoes are particularly popular among farmers in Uganda because of their flexibility in mixed cropping, because they keep well in the ground until needed, and, in the case of cassava, because little labour is involved. Cassava is currently attacked by three major diseases and pests: cassava mosaic disease, cassava bacterial blight, and the cassava green mite. The major common disease of sweet potato is caused by viruses and the most troublesome pests are *Cylas* weevils and sweet-potato caterpillar (*Acrea acerata*). Yams and cocoyams are also grown in the country, and, although no major research has been conducted on these crops, evaluation programs have been established. Future programs on root and tuber crops are aimed at: popularizing current research results among the rural communities, multiplying cassava and sweet-potato planting materials for distribution to farmers, intensifying screening and adaptation trials, and testing the performance of new varieties for adaptation to drier and semi-arid areas.

Parmi les principaux tubercules alimentaires, les cultivateurs ougandais préfèrent le manioc et la patate douce vu leur flexibilité en culture intercalaire, leur bonne conservation dans le sol jusqu'au moment où on en a besoin et, dans le cas du manioc, le peu de travail qu'il exige. Le manioc est actuellement surtout atteint par deux maladies, la mosaïque et la brûlure bactérienne, et un insecte, la teigne du manioc. Les patates douces souffrent surtout de viroses et sont attaquées par les charançons *Cylas* et les chenilles de la patate douce (*Acrea acerata*). On cultive aussi l'igname et le taro en Ouganda et même si l'on n'a fait aucune recherche importante sur leur production, des programmes d'évaluation ont été mis sur pied. Les programmes futurs sur les racines et tubercules visent à diffuser les résultats des recherches actuelles dans les communautés rurales, multiplier le matériel de plantation du manioc et de la patate douce pour le distribuer aux cultivateurs, intensifier le processus de sélection et d'adaptation et vérifier le rendement des nouvelles variétés dans des régions plus sèches et semi-arides.

In the past, farming in Uganda was strictly divided into cash and food crops, but the difference between the two is gradually disappearing. Food crops have slowly assumed cash-generating status since the country's independence. The main reasons for this change are:

- The postindependence migration of rural populations to the cities and towns, resulting in greater dependence on purchased food;
- The government policy to encourage diversification of cash crops;
- The shift of emphasis from agriculture to commerce and industry, with more food crops being produced for market and more people depending on markets as a source of food; as a result, fewer people are producing for more, and food crops have become a lucrative business; and
- The change in food habits, with increasing demand for certain food crops.

Although the number of farmers as a percentage of the population is decreasing, the production per farmer of the cash-generating food crops has increased. The food crops on which emphasis is laid are sweet potatoes, cassava, plantain, finger millet, maize, and beans.

The position of root and tuber crops in Uganda's agriculture can be clearly understood from a look at the food habits of the people. There are two distinct communities — urban and rural. The urban communities in the central, southern, and western regions depend largely on plantain, maize meal, sweet potatoes, and cassava, whereas in the urban areas of the eastern and northern re-

[1] Kawanda Research Station, Ministry of Agriculture, P.O. Box 7065, Kampala, Uganda.

gions, finger millet, sweet potato, and cassava form the major part of the diet, with a growing dependence on maize meal.

In the rural communities, finger millet, sweet potato, and cassava are, exclusively, the major food crops in the north and east, and sweet potato, cassava, and, to some extent, millet are predominant in the western rural areas. The exceptions are isolated parts of Toro, Ankole, and Kigezi where plantain, Irish potato, and sorghum are common. The central region, known as Buganda, is predominantly dependent on plantain and sweet potato but cassava is popular. Thus, sweet potato and cassava assume a very important role in the diet of Ugandans in both rural and urban communities.

Yams and cocoyams are grown on small plots, mainly in the central region. Cocoyams are also grown in banana (plantain) and coffee plantations. Some work was initiated to evaluate yam as a food crop as well as a source for medicinal preparations, and a number of progressive farmers developed interest in growing them for overseas markets where research is advanced in the extraction of diosgenin and compounds for the treatment of circulatory disorders.

Roots and tubers are likely to assume greater economic importance in Uganda for several reasons:

- The emphasis placed on the development of crops suited to the marginal and drier areas in north and northeast Uganda;
- The rehabilitation of the starch industry;
- The growing demand for sweet potato and cassava on the local market;
- The demand for cheap local feed for swine (a growing livestock industry); and
- The crop-zoning program initiated by the government to improve efficiency in production areas.

Also, there have been proposals that a factory be established to freeze cassava and sweet potato and that a research program be launched to investigate additional uses for cocoyam, which may prove to be a suitable crop for establishment in natural catchments and along rivers and flooded areas.

The statistics of yield, areas, and other production factors vary greatly. There is no reliable source, although estimates have been published by FAO (1976) and the government (Uganda, Ministry of Planning and Economic Development 1979). According to FAO, 5.3×10^5 ha were under cassava in 1976 with a per-person production of 246 kg. The same source quotes 1.41×10^5 ha being under

Table 1. Production parameters and production problems of root and tuber crops in Uganda.

	Total production (10^3 t)	Area cultivated (10^3 ha)	Yield (kg/ha)	Major production problems	Control measures
Roots/tuber	**2144**	**542**	**3958**	–	–
Cassava	1100	360	3056	Cassava green mite, cassava bacterial blight, cassava mosaic disease	Integrated pest management,[c] host-plant resistance, biological control, planting-date control
Sweet potato	660	137	4818	*A. acerata*, viruses, *Cylas* spp.	Chemicals[d], clean planting materials
Yam	na[a]	na	3110[b]	Scarabs	–
Cocoyam	na	na	2930[b]	–	–

[a] na = not available.
[b] Estimates from experimental plots.
[c] Integrated pest management: development of resistant varieties, biological control, and early-maturing varieties planted at an appropriate time.
[d] Chemicals used to control *Cylas* and *A. acerata* include carbaryl, endosulfan, sumithion, and diazinon.

sweet potatoes during the period, with the total production being 6.64 $\times$ 10^5 t of fresh tubers. No figures are available for yams and cocoyams. The estimates for 1977 are lower (Table 1).

Cassava

Generally, research on all the root crops has been limited in Uganda, and cassava is no exception. Thus, optimum agronomic practices are not known, and even those that are normally considered good are not advocated. The current practice is to plant two 25–30-cm cassava stakes crossed or a single stake laid flat 15–23 cm below the soil surface. There is no specific planting time, although traditionally farmers plant most crops during the rainy seasons (March–May and September–October). The common spacing is about 1.2 m between plants but in many cases spacings range between 1 m and 2 m.

Attempts have been made to establish trials to assess the performance of drought-resistant crops and to develop cultivars adapted to the semi-arid areas in north and northeast Uganda. Cassava is a hardy plant with a marked degree of drought endurance. Once established, the crop survives hot, dry seasons by shedding leaves and resumes growth at the return of the rains. Observation trials have shown that cassava performs well in all areas of Uganda. Further trials are needed to establish the optimum conditions for cassava in the various areas because of the variation in the climate and rainfall patterns.

Fertilizer trials have shown that cassava responds well to potash in selected areas, but high fertility results in excessive vegetative growth and thin tubers. One of the important areas of research is to determine how to minimize vegetative growth and increase the tuber number and size.

In rural areas, cassava is favoured because it can be grown with a minimum of labour, the tubers keep well in the ground until needed, it can be interplanted (mainly with beans, groundnuts, sorghum, maize, and cowpeas), its leaves can be used in vegetable soup, and the stems can be used as firewood. It is also favoured for planting on continually cultivated land of low fertility, which is common in rural areas. Because of these qualities, cassava is likely to continue to play an important role in local diets.

Production problems and current research

The main production problems for cassava in Uganda are pests and diseases: the cassava green mite (CGM); cassava mosaic disease (CMD); and cassava bacterial blight (CBB). Minor pests and diseases such as defoliators, scales, and *Cercospora* are not considered to be major factors in reduction of yield.

The biology and ecology of CGM have been extensively studied, and a country-wide survey of CBB was carried out. These studies led to the current research programs, aimed at integrated control of diseases and pests as well as improvement of yields. The programs encompass breeding and selection for resistance against CGM; study of the ecology of mites in various agronomic practices; yield performance and growth characteristics in Uganda of the IITA and other cultivars; breeding and selection against CBB; investigations of the relationships between severity of CBB, soil conditions, and meteorological factors; and assessment of root quality in CBB-infected crops.

Future programs on cassava will include: the development and screening of cultivars for drought endurance in semi-arid and dry areas of Uganda; studies of the performance of cassava in a mixed-crop system; and investigations of control of cassava mites and CBB.

Sweet potato

Relatively little crop-improvement research has been conducted on sweet potato, although its importance in local diets and its demand have greatly increased in the last 10 years. The crop is grown all over Uganda but performs better on sandy-loam soils. It is traditionally planted on rounded mounds of 45–60 cm or on ridges of similar height. Apical cuttings, the vines, about 30–45 cm long, with more than five nodes are used. Three-fourths of the cutting is buried, and the common practice is to plant three or four vines on a mound or two vines opposite each other on ridges. The spacing between mounds or ridges is about 60–76 cm. When grown on ridges, the plants along the ridge are about 30–45 cm apart. Although the planting time varies, most farmers plant during the onset of the rains, planting beans on mounds or ridges at the same time as the vines. The crop is usually weeded once, after which it produces enough vegetative cover to suppress weeds. Sweet potato is especially effective in suppressing couch grass (*Digitaria scalarum*). The crop is harvested after 3–4 months. Trials in Uganda have shown that yield is improved by the application of calcium and magnesium fertilizers.

Once established, sweet potato, like cassava, has the advantage of withstanding drought. It is a short-duration crop that requires minimum weeding, and the tubers store well in the field until required. It is a valuable crop in areas of severe drought.

Production problems and current research

Although sweet potato is frequently attacked by defoliators, the main production problems in Uganda are virus diseases, the sweet-potato caterpillar (*Acrea acerata*), the sweet-potato weevil (*Cylas* spp.), and rodents. *Acrea acerata* is effectively controlled by use of insecticides, but no effective control has been evolved against the viruses, *Cylas* weevils, or rodents. There is also a factor responsible for bitter tubers that is not well understood.

Present research programs are aimed at breeding and selection against the virus diseases, which, unfortunately, are not well understood; at providing an inventory and taxonomy of the varieties of the sweet potato; and at determining the nutrient requirements of sweet potato and the cause of bitter tubers. Future programs are planned for developing and testing varieties that are drought resistant; mixed-cropping systems that incorporate locally grown crops; and control methods for the sweet-potato caterpillars and weevils; focus will also be on monitoring increases in economic importance of pests and examining the agronomic and fertilizer requirements of sweet potato.

Sweet potato, like cassava, is regarded as an important crop and is considered to be critical to the development of subsistence agriculture in the rural communities.

Yam and cocoyam

Yams (*Dioscorea* spp.) are used as food on a limited scale. Little active research has been invested in their improvement. However, tribes in Uganda use yam and cocoyam in their diets and for medicine, possibly more than is known. No accurate figures on production and consumption are available. However, the consumption of cocoyams has increased in the urban areas, and the price of cocoyam is competitive with, or is more than, that of cassava in towns and cities like Kampala.

Some organized work was conducted by the Faculty of Agriculture, Makerere University, to evaluate the growth performance of yams. The work was discontinued probably because of lack of support. However, in the renewed interest to develop and evaluate crops for the dry parts of the country, yams will be assessed. Some of the cultivars of yams in Uganda are known to perform well during dry seasons in the savanna areas.

Cocoyams, unlike yams, are more suitable for a tropical, humid climate. The leaves and stems of some varieties are used as spinach or for vegetable soups. They are intergrown as nurse crops in plantain gardens and coffee and cocoa estates. Cocoyams have been known to withstand low rainfall conditions and, therefore, are considered to have potential for the marginal areas, although generally they are grown as rainfed crops and thrive better under warm, humid conditions. Little is known about their performance in the northern and northeastern zones of Uganda.

Problems of production and future research

Scarab beetle larvae and defoliators are known to attack yams. An anthracnose type of disease and leaf spots, possibly caused by *Cercospora* spp., are also common on yam. A mosaic-like virus disease has been observed on the crop in the central region. Similarly, cocoyams have been observed to have pests and diseases considered minor in status.

No serious research has been conducted on the effect of pests and diseases on the yield and performance of yam and cocoyam. It is planned to evaluate the traditional and imported yam varieties and to formalize the evaluation of the local cocoyam varieties. Emphasis will be placed on their food value. The proposed program will include identification and damage assessment of the various pests and diseases.

Discussion

The priority for almost all Third World countries is to improve and increase agricultural productivity at the grassroots level to achieve self-sufficiency in food of the right kind. In Uganda, more than 82.5% of the population is agricultural, and almost all of the people making up this percentage live in rural areas under subsistence agriculture. The rural population represents an important vehicle for transforming agriculture from subsistence to economic farming. It is only fair that the greatest effort be concentrated in outreach projects to educate farmers on how to produce adequate food crops compatible with their food habits while improving their productivity of the crops essential for economic development.

The available information from national research institutions and from the international agricultural research centres indicates that Uganda should embark on a program aimed at, in order of priority:

- Popularizing the current relevant research results from both the national and the international research centres among the rural communities and encouraging them to apply the results;
- Multiplying cassava and sweet-potato planting materials for distribution to farmers;
- Intensifying screening and adaptation trials at the agricultural research centres and district farm institutes, with particular reference to yield performance and resistance to CGM, CMD, and CBB; and
- Testing the performance of new varieties for adaptation to dry and semi-arid areas.

Although this document in no way presents a national policy, the contents lay a foundation for suggestions to establish an infrastructure for promoting the level of local food production needed for self-sufficiency. It also indicates the role and potential of root and tuber crops.

Permission to present this paper has been requested from the Commissioner for Agriculture, Uganda.

Malawi

R.F. Nembozanga Sauti[1]

Cassava (*Manihot esculenta*) and sweet potato (*Ipomoea batatas*) are the two commonly grown root crops in Malawi. There is also cultivation of yams and cocoyams; however, their hectarage is unknown. Information on the country's root crops is scanty, and the current research program only covers cassava and sweet potato.

Le manioc (*Manihot esculenta*) et la patate douce (*Ipomoea batatas*) sont les deux tubercules alimentaires communément cultivées au Malawi. On cultive aussi l'igname et le taro mais on ignore les superficies consacrées à ces cultures. On ne dispose que de données fragmentaires sur les cultures de racines alimentaires et les programmes de recherches en cours ne couvrent que le manioc et la patate douce.

Malawi has a population of 6 million people and a land area of 94×10^3 km^2. Maize, which is the staple food, occupies 1.4×10^6 ha. Jacoby (1967) reported that cassava was grown on 7.5×10^4 ha and sweet potato on 4.3×10^4 ha.

Cassava is the staple-food crop for 15% of the population, but the area devoted to both cassava and sweet potato has been declining. This reduction is suggested to have been the result of increased interest in crop diversification by farmers in the major cassava-growing areas and the gradual change in the eating habits of the people.

It is difficult to assess the hectarage under yams and cocoyams, but these crops are grown in some parts of the country, such as the Zomba district in the south and the Nkhata Bay district in the north.

No information is available on the sweet-potato varieties being grown, but, from a primary survey, I found more than 300 varieties of cassava. Undoubtedly, some of these varieties are similar, but their names have been changed as they were moved from one place to another.

Cassava is reported to have been introduced to Malawi between the 17th and 19th centuries (Chapola 1980) by traders and people migrating between Malawi and its neighbouring countries. The cultivation of cassava is mostly concentrated along the lakeshore districts of Mangochi, Salima, Nkhota-kota, Nkhata Bay, and Rumphi and the districts of Karonga, Mulanjo, Thyolo, Blantyre, Chiradzulu, Zomba, and Machina along the Shire highlands. Cassava is mainly grown as the staple food along the lakeshore area, although there is some cultivation of maize and rice. In contrast, it is principally grown as a famine reserve and is mostly eaten as a snack along the Shire highlands.

There are two groups of cassava cultivars being grown: bitter and sweet. The bitter cultivars are predominantly grown along the lakeshore areas, whereas the sweet types are common along the Shire highlands. Heys (n.d.) remarked that these areas could be the best sources of sweet cassava germ plasm in Africa.

Production problems

Many problems limit cassava production — among them, the absence of high-yielding varieties, the damage from diseases and pests, the lack of adequate techniques for storage, the shortage of extension services, and the poor techniques in management. Of these, disease is the major constraint, particularly cassava mosaic disease (CMD), which is primarily spread by farmers' continually planting already infected materials. Chapola (1980) and Terry and Hahn (1980) have reported losses in yields of greater than 80%

[1] Ministry of Agriculture and Natural Resources, Lunyangwa Agricultural Research Station, P.O. Box 59, Mzuzu, Malawi.

caused by the use of infected material or the use of highly susceptible varieties. In 1974, CMD was very serious in Rumphi (Chiweta/Mlowe), and, since that time, the disease has been endemic along the whole of the lakeshore area. Yields are very low (about 3 t/ha), especially where farmers plant contaminated planting materials.

The average national yield is 7–10 t/ha. However, FAO quoted average yields of 28.8 t/ha in 1970, the figure likely coming from the results found on research plots.

Bacterial blight and other serious pests, such as cassava green mite, reported in neighbouring countries, have not been found in Malawi. However, grasshoppers (*Zonocerus elegans*) and cassava scales (*Aonidomytilus* spp.) have been found to cause damage in isolated cases (A.D. Gadabu, unpublished citrus- and cassava-pest survey). Whitefly (*Bemisia tabaci*) is the only insect of major concern as a vector of CMD.

Traditionally, cassava has been regarded as a crop that requires minimum management, if any at all. It is usually pushed on to the marginal soils or is the last in the rotation. Cropping combinations involving cassava under various ecological conditions in Malawi include: imperfectly developed monoculture, *Phaseolus* beans/sweet potatoes/broad beans/cassava; maize/sorghum/cassava; pigeon peas/maize/cassava; groundnuts/cassava; etc. The time of planting depends on the objectives of the farmer. In general, maize/cassava, pigeon peas/cassava, or sorghum/maize/cassava mixtures are common along the Shire highlands, and the rest are common in parts of the lakeshore area. Often, in the north, cassava is planted first and later followed by broad beans or groundnuts and, in the south and central regions, it follows maize or pigeon peas.

In Nkhota-kota and Nkhata Bay districts, cassava is planted throughout the year because these two districts have prolonged rains. The growing season in Malawi starts in October–November in the south and in November–December in the north and ends in April–May. Hence, planting cassava with the first planting rains or planting within the first 2–3 weeks after the first rains is considered early planting. Any planting of cassava 1 month after the first planting rains is considered late planting.

In a time-of-planting trial (Table 1), statistically significant ($P < 0.001$) differences were found when planting was delayed, and the interaction between variety and time of planting was also significant. Although the results represent the findings for only 1 year and cannot be the basis for valid conclusions, they do suggest that delayed planting reduces yields irrespective of variety. Similar results have been reported by Ezumah and Okigbo (1980). The higher yields obtained from crops planted early reflect the fact that they are exposed to more months of rain.

Because of storage problems associated with cassava, farmers normally harvest what they want and leave the rest in the field until needed. Therefore, cassava commonly stands in the field for 15–36 months, depending on the variety planted. Koloweke, for example, is capable of being in the field for as long as 36 months without deteriorating, whereas Gomani is harvested between 9 and 15 months because it deteriorates very quickly thereafter. Where Koloweke is planted, much of the land remains under one crop whose total yield may not surpass that obtained from Gomani during three growing seasons.

Marketing

No cassava goes through a marketing system to consumers along the lakeshore areas, but the majority of the cassava grown along

Table 1. Fresh weight yield (t/ha) of different cassava varieties planted at different times (Baka 1978–79).[a]

Time[b]	Chitembwere	Mbundumali	Gomani	Mean
T_1	48.2	24.8	19.8	30.9
T_2	19.1	21.2	18.1	19.4
T_3	8.8	8.3	8.7	8.6
T_4	8.8	9.2	1.4	9.8
Mean	21.3	15.8	14.5	–

[a] Standard error (SE) ± var. = 1.6; SE ± T = 1.8; SE ± V × T = 3.1; site mean = 1.7.

[b] T_1 = planted 9/2/78, harvested 12/3/79; T_2 = planted 13/3/78, harvested 19/3/79; T_3 = planted 11/4/78, harvested 7/4/79; T_4 = planted 10/5/78, harvested 18/5/79.

the Shire highlands is unprocessed when it reaches the consumer through traders.

Both wholesale and retail prices are determined by supply and demand. The lowest price recorded at Ngabu, in the Shire valley was K0.02/kg (1 Kwacha = US $1.24) and the highest price recorded at Mangochi was K0.26/kg.[2] Apart from fresh cassava sales, the Mulanje, Thyolo, and Machinga (Kawinga) areas produce dried cassava chips. In 1974, the Agricultural Development and Marketing Co-operation (ADMARC) bought 8.1×10^4 t of dried chips in these areas (Chitsimo, personal communication). However, as the demand for fresh cassava has increased, the volume of sales of chips has dropped, to 1.7×10^3 t in 1979, despite ADMARC's interest in buying them.

Research activities

In Malawi, little research work appears to have been done before the early 1950s, although information on cassava contained in the annual reports of the Department of Agriculture dates back to 1935 (Davis 1976). These annual reports do not give sufficient information, and yields were rarely reported because of the problems faced during the growing period of the crop. In 1974, the Agricultural Research Department of the Ministry of Agriculture and Natural Resources started looking into cassava research very critically. Unfortunately, this effort was hampered by staffing problems. The need to do cassava research became more obvious when CMD broke out in Rumphi (Chiweta/Mlowe).

During the initial work, three varieties were chosen on the basis of popularity and not on the basis of agronomic characteristics or disease tolerance. Gomani, which is a slightly lower yielder than the other two varieties (18.8 t/ha vs 29.9 t/ha for Chitembwere and 26.7 t/ha for Mbundumali) was found to be tolerant to CMD. The Agricultural Development Department was, therefore, advised to multiply clean material of Gomani and make it available free of charge to farmers in the Chiweta/Mlowe areas. The yield differences between the varieties are not significant. Mbundumali is highly susceptible to CMD but a very popular sweet variety in Karonga.

[2] Based on biweekly market survey by staff of evaluation units and agroeconomic surveys.

In 1978, 18 cassava clones and 30 sweet-potato clones were collected and grown for agronomic and pathological evaluation. The aim of this program has been to identify high-yielding varieties within the varieties of cassava and sweet potato being grown in the country; to identify cassava varieties resistant or tolerant to CMD; and to study the agronomic characteristics of the cassava varieties for future incorporation into cropping systems suitable for smallholders.

A program to erect an ethanol plant is planned. Research is, therefore, required to provide suitable varieties of cassava and agronomic packages for the estates and individual farmers who supply the plant.

This challenging task is reflected in proposals for the 1980–81 work program of the Ministry of Agriculture and Natural Resources, which calls for:

- A second national collection of cassava based on the 1979 preliminary survey so that the commonly grown varieties from each district or ecological zone can be evaluated for various agronomic parameters and disease resistance, tolerance;
- Preliminary yield trials on 20 promising lines from 150 IITA lines undergoing initial field screening in 1979–80;
- Variety trials of 15 others, 4 of which were brought into the country in the 1950s;
- Time-of-lifting and planting trials to determine optimum periods for such operations for various promising varieties of cassava; and
- Trials for fertilizer, land preparation, spacing, and farming systems.

On sweet potato, work will include collection and evaluation of both local cultivars and crosses made in Malawi.

Conclusions

The future of root crops in Malawi, particularly cassava, cannot be overemphasized. There is an increasing interest in cassava chips, and there is no doubt that, with the need for cassava in ethanol processing, more area will be planted to the crop. Also, several national rural-development programs are including cassava in their farming systems recommendations. The increases in production need to be complemented by increased efforts in research.

Zimbabwe

A.G. Rowe[1]

The root crops cassava, sweet potato, yam, and cocoyam play a small role in the diet of the population of Zimbabwe because of the relative abundance of, and preference for, cereals such as maize and wheat. However, cassava very probably will expand into a major crop because of its potential as a source of energy and cash for the peasant farmer. It is easily and cheaply grown and is suited to the lighter soils covering much of Zimbabwe.

Les tubercules alimentaires comme le manioc, la patate douce, l'igname et le taro entrent pour peu dans le régime alimentaire des habitants du Zimbabwe, qui préfèrent le maïs et le blé, céréales relativement abondantes. Cependant, la culture du manioc va probablement prendre beaucoup d'importance car il est source d'énergie et de profits pour les cultivateurs. Le manioc se cultive bien et à peu de frais, et est adapté aux sols légers qui couvrent une bonne partie du Zimbabwe.

The cropping pattern, and particularly the use of root crops, changes as one approaches the Zambezi River area from the north. In Zimbabwe, the transition is almost complete, as the diet is derived mainly from cereal crops, principally maize but with sorghum and millet use increasing in the poor rainfall areas. The ground meal or flour prepared as a thick porridge called sadza is preferred to root crops because of its palatability and ease of storage and preparation. Recently, hybrid maize and improved production technology have dramatically increased national yields. In addition, wheat production increased to self-sufficiency in 1978, and bread consumption continues to increase and is now found even in remote rural areas.

Cassava

Cassava is a relatively recent introduction, brought in by migrant workers. These people were mainly Malawians who were employed on large commercial farms, and, as a result, most cassava in Zimbabwe is found in small plantings around houses on the farms. Spread to the tribal areas has occurred to a limited extent as Zimbabwean farm workers return to their homes. A small percentage of homes have a few plants that are used occasionally for a change in diet. Preparation is limited to boiling or roasting.

Interest in cassava has been stimulated recently by an increase in the price of maize. As in Europe, in Zimbabwe, the cereal-price increase has made intensive livestock feeding prohibitively expensive, and cassava is seen as a potential replacement feedstuff.

Similarly, cassava could replace maize in industrial starch production. A commercial company is currently offering Z$20/t of wet roots and is sponsoring plantings on commercial farms and in the tribal areas.

Zimbabwe has, at the moment, a 20% ethanol blend in its gasoline. If good yields can be achieved, cassava may compete favourably with sugarcane as a raw material for this ethanol production, particularly if the sugar price increases.

Research

To date, research work has been limited to collecting and bulking local material and introducing four Australian varieties (MAus. 1, 7, 10, 19). At present, however, several trials are under way, including:

- Growth analyses — four sites, with monthly sampling over 3 years;
- Variety trials — three sites, with local selections and the Australian imports;

[1] Department of Research and Specialist Services, P.O. Box 8100, Causeway, Harare, Zimbabwe.

- Date-of-planting trial — one site; and
- Storage of seed material — one site.

Initial results have given approximately 25 t/ha at middle- and high-altitude areas and in excess of 30 t/ha at low-altitude areas under irrigation.

Some cassava mosaic disease has been observed but no other diseases. The only pests to date are cutworms that damage the young emerging shoots and wild pigs and porcupines that damage mature roots. Moderate root rot in the winter has been experienced, particularly where frosts have occurred.

Sweet potato

Sweet potatoes have been grown for many years and form an important part of the diet when available. Their limited storage life, however, restricts consumption to a short period after the main rains.

The tubers are eaten raw, boiled, or roasted. Planting is done on ridges or mounds, and green manure (as much as is available) is incorporated.

Early lush growth is eaten by sheep and goats, and moles cause considerable damage to the tubers. Pests are not a great problem, although the sphingid caterpillars of the hawkmoth *Herse convolvuli* and sweet-potato weevil *Cylas* sp. can cause damage to the roots.

No research work has been conducted in Zimbabwe, but yields are about 30–40 t/ha on commercial farms, and the Horticultural Research Institute has recently started the production of virus-free material for distribution.

Yam and cocoyam

There is little information about yam and cocoyam in Zimbabwe. They are grown in the eastern highlands where higher rainfall is experienced and are restricted to waterlogged marshy vlei areas. Not extensively utilized, they tend to be regarded as a delicacy. There are two types, one called Majo and the other Madumbe, the former having edible leaves. No research has been conducted on these crops to date in Zimbabwe.

Ethiopia

Terefe Belehu[1]

Several million people in Ethiopia depend on root and tuber crops as a staple food; therefore, a national effort has been directed toward developing acceptable cultivars and appropriate cultural practices so that improved yields and better quality products can be obtained. The root and tuber crops given priority in this program are potato, sweet potato, enset, yam, taro, anchote, and cassava.

Plusieurs millions d'Éthiopiens dépendent des racines et des tubercules alimentaires pour leur subsistance; on a donc mis en oeuvre un programme national de développement de cultivars acceptables et de méthodes agricoles adéquates pour obtenir des rendements accrus et des produits de meilleure qualité. Dans le cadre de ce programme, on accorde la priorité à la pomme de terre, à la patate douce, à l'«enset», à l'igname, au taro, à l'«anchote» et au manioc.

At present, most of the people in Ethiopia depend on cereal crops as their primary staple food; however, there are probably several million who rely on root and tuber crops. The regions where the latter crops are cultivated as sources of food are the most densely populated areas in the country — hence, the areas where land-use efficiency and crop productivity are particularly important. This is one reason for concerted national effort toward the improvement of root and tuber crops.

For the last few years, roots and tubers have been studied under the auspices of the state Horticulture Department and have been the focus of one of the research teams. The objective of this team is to develop cultivars and cultural practices that can give improved yields and better quality products. The root and tuber crops given priority by the team at present include potato, sweet potato, enset, yam, taro, anchote, and cassava.

Enset (*Enset ventricosa*) is known as a food crop only in Ethiopia and is sometimes referred to as false banana because of its appearance. It is grown mainly in the south and southwestern part of the country. About 7 million people (i.e., one-quarter of the population) depend on enset as a source of food and fibre. It is peeled, pulverized, and allowed to ferment. The dough is served in different ways.

At present, research on this crop is focusing on the development of cultivars resistant to bacterial wilt (*Xanthomonas musacearum*), which threatens production in the country. Other problems being addressed are the long growing season, which can be up to 8 years at the farmer level, and the difficulties inherent in the method of processing.

Potato (*Solanum tuberosum*) is a popular vegetable throughout the country, although its production in many areas has been constrained by late blight (*Phytophthora infestans*). Thus, a project is being launched to develop cultivars that are resistant to this and other common potato diseases. The results to date have been promising, and it is hoped that the cultivars will be ready for release in the coming years.

Sweet potato (*Ipomoea batatas*) is eaten primarily in the southern and eastern parts of Ethiopia. The roots are boiled and served as a vegetable. In the eastern part of Ethiopia, the leaves and stems of the crop are used as a feedstuff for cattle.

In the past few years, an attempt has been made to collect cultivars both locally and from abroad and to evaluate them under different climatic conditions. The most promising have been studied thoroughly and released to farmers. Acceptance, even by nomads, has been high, and sweet-potato production has spread

[1] Institute of Agricultural Research, P.O. Box 2003, Addis Ababa, Ethiopia.

quickly into regions where it was unknown a few years ago.

At present, screening of cultivars and studies of optimum growing conditions (spacing, planting time, harvesting time, fertilization, method of planting, etc.) are being conducted at different stations. Intercropping sweet potato with sorghum and corn is of interest, and the selection of cultivars that are compatible for intercropping has been given priority.

Although satisfactory results have been obtained on sweet-potato trials, attention should be given to:

- Development of cultivars that are adaptable to different regions, particularly arid ones;
- Improvement of storage systems, with the aim being to find methods that reduce losses and can be used by smallholders; and
- Study of a stem blight (*Alternaria* sp.) and a leaf minor (*Bedellia somnulentella*), which have recently become important. Sweet-potato weevil (*Cylas compressus*) is another pest problem but is not serious in Ethiopia at present.

Other minor root and tuber crops are important in some areas, mainly south and southwestern Ethiopia, but are given less research emphasis because of a shortage of trained personnel. These include taro (*Colocasia esculenta*), yam (*Dioscorea alata*), cassava (*Manihot esculenta*), anchote (*Coccinia abyssinica*), and wolaita denech. The last two crops are indigenous to Ethiopia, and, as is the case for enset, no research results on these crops are expected from other countries.

Burundi

D. Cimpaye[1]

Cassava and sweet potatoes are the most important root crops grown in Burundi; cush-cush and yams are also grown. Current production levels and the constraints to production are outlined, the main constraints being land availability, trained personnel, soil fertility, and diseases and pests. Plans are under way to establish a research program on the crops.

Le manioc et la patate douce sont les principaux tubercules alimentaires cultivés au Burundi, où l'on retrouve aussi le taro et l'igname. On donne un aperçu des niveaux de production actuels et des obstacles rencontrés, les principaux étant le manque de terres et leur peu de fertilité, la pénurie de main-d'oeuvre qualifiée, les maladies et insectes. On planifie en ce moment un programme de recherche sur la production des cultures.

Cassava and sweet potatoes are of considerable importance in Burundi where they occupy more than 30% of the total planted area every year; are the famine food crops for the period between the two growing seasons (October–December and February–May) when rainfall is fairly irregular; and are generally grown on poor, leached soils that are unsuitable for crops such as legumes, grains, and bananas.

Cassava is grown at an altitude of 800 m in the Ruzizi plain and in the coastal plain of Lake Tanganyika up to an altitude of 1800 m. This crop is vital in the low-lying areas (800–1300 m) because, together with fish, it constitutes the staple food of the people.

Sweet potatoes are grown at altitudes up to 2000 m, either on the hills during the rainy season or in the fairly numerous valleys during the dry season.

Cush-cush (*Colocasia*) and yams are less important and constitute what is known as homestead crops. Yams are only found in the Gitega and Bukirasazi administrative districts, which, together, form the natural region known as Kirimiro, with an altitude of about 1700 m. Most families in this region have a few yam plants along the fences around their homes. Quite often, some cush-cush plants, usually no more than 50, are found growing in the banana plantation that is indispensable to every family. Cush-cush is mostly found in Kirimiro (centre of the country) and in Mosso (eastern savanna).

The annual reports published by the Department of Agronomy for the 1976–79 cropping seasons provide figures on the area devoted to these crops and their levels of production (Table 1). The differences in area as well as in production between sweet potatoes and cush-cush during the 1976–77 and the 1977–78 seasons are mainly attributable to a lack of rain and early infestation of sweet potatoes by caterpillars. There is no record of the amount of sweet potatoes destroyed.

Figures from the 1978–79 crop season indicate that root and tuber crops occupied 6.36 ×

[1] Institut des sciences agronomiques du Burundi (ISABU), B.P. 795, Bujumbura, Burundi.

Table 1. Production (10^3 t) and area (10^3 ha) devoted to root and tuber crops in Burundi.

	1976–77		1977–78		1978–79	
	Area	Production	Area	Production	Area	Production
Cassava	363.9	3814.9	276.8	4318.8	398.1	4492.1
Sweet potato	207.5	2199.1	198.7	2735.1	196.8	1787.8
Cush-cush	65.5	510.3	32.5	238.7	39.0	317.3
Yams	0.9	17.0	0.3	4.8	2.5	13.9

10^5 ha, or 36.5% of the total planted area (1.74 × 10^6 ha), 22.9% of the total area being devoted to cassava, 11.3% to sweet potatoes, 2.2% to cush-cush; and 0.1% to yams. The importance of these crops lies in the fact that they enable the people to live through the dry season, even though their quantitative importance is relatively small compared with that for other crops.

Limits to production

Because of the high population-growth rate in Burundi, land is becoming increasingly scarce, and even the eastern savanna, which was previously considered unsuitable for development, is now being utilized within the framework of an integrated program.

The government is considering the reallocation of land throughout the country and may restrict each family to tenure of not more than 5 ha. Root and tuber crops, which are grown far from the home, will be affected by this reduction in available land because they do not fit easily into the traditional crop rotation and will, therefore, tend to be replaced by crops such as beans, corn, and wheat.

Labour is also becoming scarce in the rural areas, particularly in Mosso and in the coastal plain of Lake Tanganyika where rural-development projects are absorbing all the available workers.

This labour constraint may be resolved by a national authority responsible for the mechanization of agriculture, which was set up nearly 2 years ago. This authority now operates well in the low-lying region where the population is organized in villages, but it is likely to be much less effective in the hinterland.

Animal power is being used on a limited scale in the Imbo plain, introduced relatively recently by the Société régionale de développement de l'Imbo; however, this method of farming is also likely to be unsuitable in the hinterland because the people in this area venerate cattle.

One of the major factors limiting cassava yields is that this crop is grown only on poor and exhausted soils. For example, cassava is frequently found growing in completely leached soils on hillsides. In contrast, sweet potatoes are usually grown on lowlands, which tend to be reasonably fertile.

The diseases and pests attacking these two root crops cause less damage in Burundi than has been reported elsewhere. Cassava suffers somewhat from the cassava green mite, cassava mosaic, cercosporiosis, and the mealybug; however, cassava bacterial blight has not yet been reported in the country, and there is a plant inspection team from the Institut des sciences agronomiques du Burundi (ISABU) that operates in the rural areas in the vicinity of ISABU stations.

The major pest of sweet potato in Burundi is the defoliating caterpillar, and the Depart-

Table 2. Fresh weight yield (t/ha) of IITA varieties of cassava compared with local varieties at Mosso and Imbo.

Variety	Mosso 1978–79	Mosso 1979–80	Imbo (1979–80 8–9 months)
IITA			
TMS 30555	20	22	34
30572	23	22	30
31042	12	20	30
30741	15	22	27
30404	14	21	–
30728/3	–	15	31
31136	–	14	30
31034	11	17	26
30728/2	11	16	18
30728/1	24	19	10
30983	12	13	20
30568	13	13	11
Local			
Creolinha	37	29	–
Nakarasi	–	–	50

Table 3. Fresh weight yield (t/ha) of IITA varieties of sweet potato compared with local variety grown at Mosso.

Variety	Yield (t/ha)
IITA	
TIS 2498	22.1
TIS 2532	21.5
TIS 2525	12.8
TIS 3030	11.9
TIB 3	10.3
TIS 5016/1	8.1
TIS 2330	5.9
TIS 5016/2	4.4
TIS 5192	1.3
Local	
Caroline Lee	15.8

ment of Agronomy has countered the problem by setting up teams to undertake treatment of all areas of the country where sweet potatoes are grown.

Research

Research on these crops has not yet started in Burundi. The reason is that, before the country's independence, INEAC (National Institute for the Economic Study of the Congo) activities covered three countries (Rwanda, Burundi, and Zaire), and material produced for one area in Rwanda or Zaire was used in a similar area in Burundi. Thus, the cassava material selected or introduced by INEAC (Amer de 6 mois, Eala 07, and Creolinha-Yava) is still being widely used.

Collections of cassava and sweet-potato varieties have been built up with plants from the rural areas and have been supplemented by IITA varieties. Some of the local varieties compare favourably with the IITA imports (Tables 2 and 3).

A research program is due to start in January 1981, and steps will be taken to establish cooperation with IITA. The Institut de recherches agronomique et zootechnique (IRAZ) has just been set up by the Communauté des pays des grands lacs, and a joint program will be drawn up. It is expected that researchers at this new institute will devote some attention to the problems of root and tuber crops.

Zaire

T.P. Singh and N.B. Lutaladio[1]

Cassava is the most important root crop grown in Zaire, and both the roots and leaves are consumed. Root yields from local varieties are low because of their low genetic potential and susceptibility to major pests and diseases. Research for the improvement of cassava in Zaire was initiated in 1976, and breeding efforts have resulted in the identification and development of high-yielding varieties with field resistance to major diseases under different soil conditions. A source of resistance to mealybug has been identified in both wild and cultivated cassava, and resistance to cassava green mite is available. Cassava germ-plasm screening for leaf HCN content has been started. Agronomic studies have resulted in recommendations to improve cassava production and income at the farm level. A strategy to diffuse improved cassava varieties and cultural practices for adoption by farmers in the absence of an effective national extension service has been suggested.

Le manioc est la principale racine alimentaire cultivée au Zaïre, où l'on consomme la racine et les feuilles. Les variétés locales n'ont qu'un faible rendement vu leur faible potentiel génétique et leur sensibilité aux insectes et aux maladies. On a entrepris des recherches visant à l'amélioration du manioc zaïrois en 1976 et les travaux de sélection ont permis d'identifier et de développer des variétés à rendement élevé, résistant bien aux principales maladies dans divers types de sols. On a identifié une source de résistance à la cochenille dans le manioc sauvage et cultivé et on dispose aussi d'un agent de résistance à la teigne. On a commencé la sélection du matériel génétique du manioc en fonction du contenu en HCN dans les feuilles. Les études agronomiques ont abouti à des recommandations sur l'amélioration de la production du manioc et de sa rentabilité au niveau de la ferme. En l'absence d'un service national de vulgarisation, on a proposé une stratégie de diffusion des variétés et pratiques agricoles améliorées auprès des agriculteurs.

The five major root crops cultivated in Zaire are cassava, sweet potato, yams, taro, and potato. However, cassava occupies 94.4% of the total area under root crops and accounts for 95.3% of the root crops produced in Zaire. Cassava is grown in all regions of the country under various soil and ecological conditions. The regions of Kivu and Bandundu are the foremost in terms of area and production of cassava (Table 1). Over the past 5 years, the national average cassava yield varied between 6.62 t/ha and 7.02 t/ha. Of the different regions of Zaire, Kasai Occidental had the highest cassava yield (7.32 t/ha), and Bas-

[1] Programme national manioc (PRONAM), B.P. 11635, Kinshasa, Zaire.

Table 1. Area, production, and yield of cassava in the different regions of Zaire (1978).

Region	Area (10^3 ha)	Production (10^3 t)	Yield (t/ha)
Kivu	299.6	2022.9	6.75
Bandundu	290.3	2088.8	7.20
Shaba	235.6	1529.0	6.49
Kasai Occidental	209.1	1530.8	7.32
Haut-Zaire	199.9	1361.6	6.81
Equateur	186.4	1222.1	6.56
Kasai Oriental	181.2	1256.9	6.94
Bas-Zaire	176.1	766.8	4.35

Source: Department of Agriculture and Rural Development 1977–78.

Zaire had the lowest (4.35 t/ha) in 1978. The extraordinarily low cassava yield in Bas-Zaire in 1978 was primarily caused by a serious drought. The average cassava yield in Bas-Zaire in previous years was about 7 t/ha.

The national average cassava yield of 6.7 t/ha is very low compared with potential yields of more than 20 t/ha from improved cultivars. The major cause for this low yield is the susceptibility of local varieties to pests and diseases. The most devastating of all the diseases is cassava bacterial blight (CBB), which has been observed throughout the country and is capable of causing average yield losses of 57–90% in susceptible cassava varieties (Terry 1978b). In addition to losses in root yield, CBB causes serious reductions in leaf production because of leaf wilting and stem dieback. An average root-yield reduction of 50% from mosaic infection has been reported (Jennings 1970). Anthracnose, a fungal disease, also causes considerable root-yield reductions, although no precise estimate is available.

The losses caused by diseases are compounded by those caused by attacks from mealybugs and green mites in the dry season. Root-yield reductions caused by mealybugs vary between 54% and 85%, depending on plant age at the time of infestation. Yield losses from green-mite infestation are not known precisely.

In 1974, because of the seriousness of CBB in Zaire and its effect on cassava production, the government asked for technical assistance from the International Institute of Tropical Agriculture (IITA) to conduct research into the improvement of cassava. As a result, Programme national manioc (PRONAM) came into existence to:

- Identify and develop cassava varieties, with high-yield potential, suitable for the different regions of Zaire;
- Develop a package of cultural practices suitable for cassava cultivation in the different ecological areas in Zaire;
- Identify sources of resistance to major pests and diseases of cassava and incorporate the resistance genes into high-yielding varieties; and
- Develop cassava varieties with roots of acceptable quality and low levels of HCN and with protein-rich palatable leaves.

Achieving these objectives requires a multidisciplinary approach, with the agronomist, pathologist, and entomologist working together in the screening, selection, and evaluation stages and sharing information with the extensionist and personnel from national organizations who will take responsibility for passing results to the farmers.

Varietal improvement

At the beginning of the project, the locally cultivated varieties were assembled and evaluated for their disease resistance. All proved to be susceptible to CBB, mosaic, and anthracnose; therefore, other potential sources of resistance were required. As a result, large amounts of cassava seed were introduced from IITA's root and tuber program and were augmented by seeds collected from local varieties. This assembled genetic variability was screened against diseases and showed considerable variation (PRONAM 1975) (Table 2).

Research was initiated at a single station at M'vuazi, but later, with the availability of Zairois trained at IITA, PRONAM established three more stations, one each in Bandundu, Kasai, and Kivu, to cater to the needs of these areas. The seed material obtained from outside and generated at M'vuazi is distributed to these stations for the selection of agronomically superior clones resistant to the pests and diseases of the area.

Table 2. Percentage of cassava seedlings rated in disease classes 1 and 2 for CBB, mosaic, and anthracnose.

Source of seed	Plants screened	Plants in classes 1 and 2 (% of total)		
		CBB	CMD	Anthracnose
IITA B-Ext. (OP)[a]	520	54	65	42
IITA 58308 (OP)	492	65	59	47
IITA Hybr. B-Ext.	615	63	48	48
INERA (OP)	324	25	12	18
Farmers' seed	165	29	8	6

[a] OP = open pollinated.

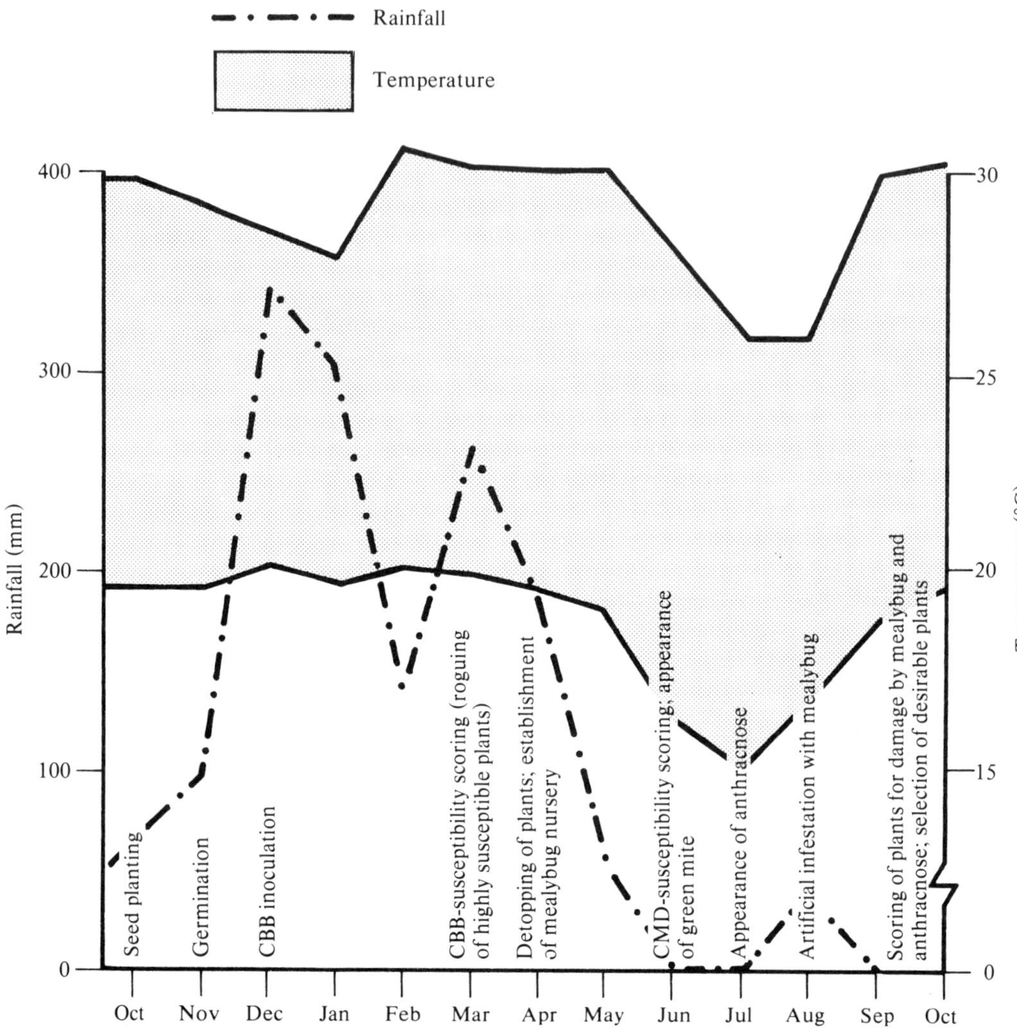

Fig. 1. *Calendar for screening of a seedling nursery in relation to rainfall and temperature pattern at M'vuazi station.*

The establishment of these substations obviates the need to move vegetative planting material from one place to another and thus checks the spread of pests and diseases; it also facilitates the adoption of improved material through selections made under appropriate environmental conditions.

At M'vuazi, between 20 000 and 100 000 seeds are planted for screening against pests and diseases. Efficient and thorough screening of this large population against all pests and diseases requires a systematic procedure. The procedure developed at M'vuazi (Singh 1979) takes into account the optimum time of the appearance of diseases or pests in relation to the climate and plant age most suitable for artificial infestation. A screening calendar (Fig. 1) can be modified to suit different planting times, climatic variations, etc. This procedure was used in the selection of 109 clones showing multiple disease resistance (mosaic, CBB, anthracnose). Earlier, a number of clones possessing a reasonable level of resistance were identified and have been evaluated for their yield potential (Table 3) in Bas-Zaire and Bandundu.

For resistance to pests, a large seedling nursery was screened under artificial conditions. Only three wild cassava plants showed resistance to mealybug, and they

Table 3. Yield (t/ha) of some improved cassava varieties compared with a local control variety in Bas-Zaire and Bandundu regions.

	Yield
Bas-Zaire (valley)	
A56	23.3
3070/4	20.7
30344/6	20.1
30179/2	20.1
02864 (local)	10.7
Bandundu (forest)	
30577/5	26.3
F150	24.1
F112	21.8
F100	19.9
Mundjoko (local)	12.6
Bandundu (plateau)	
F113	6.0
F188	5.3
A43	4.9
30741/5	4.1
Mundjoko (local)	1.5

were utilized in a hybridization program to transfer genes conferring resistance into locally adapted varieties (PRONAM 1978). Recently, in another seedling nursery screened against mealybug, some plants of cultivated cassava were identified as showing some level of resistance, and they are currently being retested in an effort to confirm the findings (PRONAM 1979). Resistance to cassava green mite has also been identified.

Cassava germ plasm is being screened qualitatively for HCN content in the leaves, and preliminary results show low HCN in only 4.1% of the clones.

Agronomy

Agronomic aspects of cassava production have been thoroughly investigated. In fact, studies conducted from 1976 to 1979 allowed some improved cultural practices to be recommended (Ezumah 1980):

- Timely planting of cassava is very important, and, under the conditions in Bas-Zaire, December–January planting gives the maximum yield.
- The use of disease-free planting material gives optimal emergence rates and minimizes disease problems.
- The optimum plant population in poor sandy soil is between 10 000 and 15 000 plants/ha, but the population can be increased to a maximum 26 000 plants/ha with certain soil conditions and cultivars.
- Cassava grown in mixtures with groundnut or *Phaseolus* beans produces higher revenue than a sole crop of cassava.
- Weeding of cassava, particularly during the first 3 months, is very critical. Three weedings done at 30, 60, and 105 days after planting are sufficient for good cassava growth. Delay in first weeding causes yield reductions of 40–60%.
- Commencing with 4-month-old plants, leaf pruning done at a 2-month interval results in a slight, favourable response in root yield. The income additional from the sale of leaves is a bonus.

Entomology and pathology

Studies have been designed to determine the factors that favour the incidence of pests and diseases and to develop quick and efficient methods of screening under artificial conditions (PRONAM 1978). The biology of cassava pests, particularly the mealybug, has been thoroughly studied (PRONAM 1977), and these studies have made an important contribution to the development of sound procedures for the screening of genetic variability.

Farmer acceptance

Adoption of improved cassava varieties and cultural practices by farmers is essential if cassava production in the country is to increase. Insufficient government effort has been directed to this important aspect, and PRONAM has directed its efforts mainly toward cassava research. In 1980, PRONAM added an extension component to the existing research setup through the financial assistance of the United States Agency for International Development and the technical assistance of IITA. The new wing of PRONAM has been charged with the responsibility of conducting multilocation yield trials, farm-level testing of a few selected varieties and cultural practices, and the multiplication and distribution of vegetative planting material of selected cassava varieties. To achieve this on a large scale in the absence of a well-organized extension agency in Zaire, PRO-

NAM has enlisted the cooperation of the Institut national pour l'étude et la recherche agronomique (INERA), Groupe économie rurale (GER), a subsidiary of the Department of Agriculture and Rural Development, community development projects, missions involved in agriculture, and intergovernment cooperative projects. In addition, cooperation from certain private organizations in the multiplication of material would be welcomed.

At present, multilocation trials are being conducted in the fields of some INERA and mission centres, and multiplication of the improved material has already begun and will be strengthened as the extension wing becomes well established.

Swaziland

W. Godfrey-Sam-Aggrey[1]

General background information encompassing the agroecological zones and agriculture of Swaziland is presented. The root and tuber crops grown in Swaziland are potato (*Solanum tuberosum*), sweet potato (*Ipomoea batatas*), cassava (*Manihot esculenta*), and taro (*Colocasia esculenta* var *esculenta*), locally called Madumba. The total areas under each crop and their locations are cited. Potato is the most important, followed by sweet potato, whereas cassava and taro are of less importance in terms of area and use. Research on these crops and future research directions are reviewed. There is potential for increased yields of potato and sweet potato and for the increased use of cassava for animal feed and for the industrial production of alcohol for fuel. In the case of Madumba, the collection, characterization, and description of available local material should be attempted so that the spectrum of available germ plasm in the country can be monitored. It is imperative for research interests to be developed in both cassava and taro because of their potential.

Présentation générale du contexte agro-écologique et agricole au Swaziland. Les racines et tubercules alimentaires cultivés au Swaziland sont la pomme de terre (*Solanum tuberosum*), la patate douce (*Ipomoea batatas*), le manioc (*Manihot esculenta*) et le taro (*Colocasia esculenta* var *esculenta*), appelé sur place Madumba. On indique la superficie et l'emplacement de chaque culture; celle de la pomme de terre vient en tête suivie par la patate douce, le manioc et le taro étant moins importants en termes de surface cultivée et de consommation. On passe en revue les recherches sur ces cultures et les orientations de recherches futures. On peut accroître les rendements de la pomme de terre et de la patate douce; le manioc peut être plus utilisé pour la consommation animale et la production industrielle d'alcool servant de carburant. Quant au Madumba, il faudrait tenter de recueillir, caractériser et décrire le matériel local pour vérifier la gamme de matériel génétique dont on dispose au pays. Il faut intensifier les recherches sur le manioc et le taro, étant donné leur potentiel de production.

The Kingdom of Swaziland is a small, landlocked country with a total population of 528 000 (1976), which is growing annually 3.2%. About 90% of the population is rural, and, of this, about 70% lives in scattered small homesteads on Swazi Nation Land, which comprises 57% of the total land area. Here, traditional authority and land tenure prevail. The remaining population lives on freehold land, most of it owned and farmed by large companies and farmers. The average population density is about 28 persons/km^2.

There are four markedly distinct topographic regions, each extending in zones from north to south: the Highveld, Middleveld, Lowveld, and the Lubombo plateau.

The Highveld (5200 km^2) is a strip on the western escarpment of the South African plateau. Its altitude ranges from 1000 m to 1800 m, and its climate is humid, with annual rainfall of between 1000 mm and 1750 mm, much of which falls between September and April. The soils are deep red and yellow ferralites on an ancient granite base. The main agricultural activity is pastoral farming, with some mixed cropping. Research sites are located at Mangcongco and Hebron.

Middleveld (4700 km^2) is a belt of foothills to the east of the Highveld that has variable topography and an altitude of between 350 m and 1000 m. The climate is subhumid, with rainfall of 750–1200 mm occurring mainly in the summer. Typically, ferralitic and ferruginous tropical soils overlie granite. Agricultural activity is a complex mixture of ranching, mixed rainfed and irrigated field crops, and horticultural crops. Research sites

[1] Department of Crop Production, University College of Swaziland, P.O. Luyengo, Swaziland (Professor of Crop Production and Senior Crop Production Officer, FAO Project GCP/SWA/014/SWE, Swaziland).

are located at Malkerns, Nhlangano, and Luve.

The Lowveld (6200 km^2) is a belt east of the Middleveld. The topography is gently undulating within altitude limits of 60–375 m. The climate is subarid, with restricted and erratic summer rainfall of 500–800 mm, and droughts are a hazard. The soils are derived from granite and Karroo sediments in the west and basic igneous rocks in the east. Problems of salinity and alkalinity are encountered in some areas. Cattle ranching and the production of sugarcane, citrus fruits, and cotton under irrigation are the main agricultural activities. Mixed cropping is also practiced. Research sites are located at Big Bend and Vuvulane.

The Lubombo plateau (1300 km^2) is a strip on the eastern border. It has a rolling topography with deep gorges and an altitude of 450–850 m. The climate is frostfree, and annual rainfall is 750–900 mm. The predominantly red and brown ferruginous, tropical soils are derived from rhyolites and basalts. The main agricultural activities are ranching and mixed cropping (Murdoch 1970).

Swaziland has a diverse agricultural sector that is differentiated by geographic regions and modernization. Of the total land area (1.7×10^6 ha), 65% is used for grazing, 10% is under cultivation, and 10% is considered suitable for intensive cultivation. Irrigation is important, and an estimated 2.5×10^4 ha are irrigated (mostly in the Lowveld). Forests (mostly plantations) cover about 6% of the land area. Swaziland is a commercial exporter of sugar, citrus fruits, cotton, forest products, and livestock but must import food for its people.

Subsistence farming is practiced on Swazi Nation Land where there are more than 39 000 families farming holdings that average fewer than 3 ha. Because production is for subsistence, less than 10% enters the market. Maize is the principal and staple-food crop of Swaziland, and it occupies about 80% of the total crop area. Other important crops are groundnuts, sorghum, beans, potatoes, sweet potatoes, and pumpkins. In recent years, small farmers have also engaged in the production of cash crops, notably cotton, tobacco, and vegetables. Mixed cropping is common. Women are the principal farmers; men are concerned with the livestock. Cultivable land is allocated to families by the chiefs. The land may not be sold; however, land may be passed from parent to child. Fragmentation of farming plots is common and is becoming a serious problem in some areas as population pressures increase.

Root crops

Potato (*Solanum tuberosum*) is the most important tuber crop and occupies 330 ha of Swazi Nation Land. The areas devoted to pure (P) and mixed (M) cropping are as follows: Highveld P 13 ha, M 49 ha; Middleveld P none, M 18 ha; Lowveld P 27 ha, M 223 ha; Lubombo plateau none. Approximately 88% of the total land area on Swazi Nation Land is devoted to mixed cropping.

Potato is rainfed in the Highveld but requires supplemental irrigation in the Middleveld and the Lowveld. Eelworms and late and early blights are serious problems. Varietal trials have not received much attention. BP/1 (110 days) and Cedara (140 days) are moderately resistant and resistant, respectively, to late blight (*Phytophthora infestans*). BP/1 certified seeds are being multiplied for distribution to farmers.

Sweet potato (*Ipomoea batatas*) is the second most important tuber crop. The total area under the crop is 1.5×10^3 ha. Of this total, 550 ha are on Swazi Nation Land, especially the Middleveld. The areas under pure (P) and mixed (M) cropping on Swazi Nation Land are as follows: Highveld P 30 ha, M 70 ha; Middleveld P 192 ha, M 153 ha; Lowveld P 31 ha, M 34 ha; and Lubombo P 39 ha, M 1 ha.

Haque et al. (1979) noted that a large number of variety-evaluation trials involving both local and imported cultivars, mainly from South Africa, had been conducted. They further pointed out that there was an absence of any record of agronomic work, including fertilizer requirements, on sweet potato in Swaziland and observed that current recommendations for the crop have been extrapolated from elsewhere, particularly South Africa.

The crop is grown in all regions in Swaziland and is either rainfed or irrigated. Hartebees (red rind, white flesh), Kudu (white rind, white flesh), and Wildebees (red rind, white flesh), which mature in 4–6 months, are commonly grown. Porto Rico (gold rind and flesh), which takes 3 months to mature, was imported in 1978 and has been multiplied at the Faculty of Agriculture, University College, Swaziland (UCS), Luyengo. Vines from seeds obtained from IITA died, in seedbed, of an unknown cause. The Chinese Agricultural Mission in Swaziland recently imported five new cultivars from Taiwan for evaluation and multiplication. The major pests are *Herse convolvuli* (leaf-eating caterpillars of the sphinx moth) and *Cylas formicarius* (sweet-potato weevil). The common diseases are mosaic and rosette.

Although cassava grows in all regions of Swaziland, it is more common in the Lowveld on small farms as an intercrop and in the Middleveld as a garden crop and hedge plants. The increasing importance of this crop for export and domestic use in most tropical and subtropical regions of the world cannot be overemphasized. Yet, little research has been conducted on the crop in Swaziland. Although it is eaten in small amounts, a taboo exists on its permanent use as a food for adults and the roots are offered only to children. The Meat Corporation of Swaziland has two fattening ranches where cassava is planted in large areas and the fresh roots are fed to pigs and cattle. Two cultivars Swazi White and M-Saf 1 ex Zululand are grown by the Meat Corporation for feed.

The Swazi White cultivar is susceptible to bacterial blight and mosaic. Nevertheless, the Meat Corporation plans to replace 25% of the maize used in the fattening feed with cassava chips, pellets, root, or meal (Tudo-Owen, personal communication 1980).

My colleagues and I in 1978–79 initiated research on agroeconomic aspects of cassava in Swaziland with funding from the International Foundation for Science in Sweden.

The research involves the collection, characterization, and evaluation of local and imported cassava cultivars to provide germ plasm for future work on adaptability, disease and pest resistance or tolerance, and yield potential. Agronomic and intercropping studies including *Phaseolus* spp., *Vigna unguiculata,* sweet potato, and pumpkins with cassava as a dominant crop will be undertaken. The research will span the four different agroecological zones, and the economics of production and market demand will be studied. Selections will be made for high leaf protein, palatability, and low HCN content. Seedlings obtained from IITA seeds were cloned in 1978 and observed. In 1979, eight elite cassava cultivars were selected from the 100 cassava cultivars introduced into Swaziland. They were coded and compared with a local cultivar on the basis of fresh- and dry-root yield. The eight elite cultivars are now being described and characterized, and studies are being conducted on bulking rates 8 months after planting under Middleveld conditions; harvest indices; the HCN content in the leaves, peel, and flesh; the starch-content profile at monthly intervals from 8 months after planting; the rapid multiplication of the elite cultivars for larger plantings; and the damage from diseases and pests. In addition, more local cultivars are being collected, characterized, and described.

Taro (Madumba) is grown in mixed-cropping systems in the Highveld, Middleveld, and Lowveld on a total of 279 ha. On Swazi Nation Land, the areas are Highveld 58 ha, Middleveld 67 ha, and Lowveld 154 ha, and the Lubombo plateau none. No research has been conducted on this crop. Taro is often called yam in Swaziland, but this is a misnomer because the species is *Colocasia esculenta* var *esculenta*.

Future research

Potato and sweet potato have been studied a great deal. Some research is also being done on cassava, but no work has been done on Madumba. Imports of improved cultivars of potato and sweet potato from international research institutes to broaden the spectrum of germ plasm for incorporation into evaluation and adaptability trials in the different agroecological zones of Swaziland are desirable. And a national coordinated root-crops improvement program should be initiated. Research work on varietal improvement, agronomy, crop protection, physiology, and economics has been suggested (Haque et al. 1979).

References

Acland, J.D. 1975. East African crops. London, England, Longmans.

Arrandeau, M. 1967. Cassava in the Malagasy Republic: research and results. Paper prepared for the First Symposium of the International Society for Tropical Root Crops, University of the West Indies, St. Augustine, Trinidad, 2–8 April 1967, 1, 180.

AVRDC (Asian Vegetable Research and Development Center). 1976. Annual report. Taiwan, AVRDC.

Bartlett, B.R. 1978. Pseudococcidae. In Clausen, C.P. ed., Introduced Parasites and Predators of Arthropod Pests and Weeds: a World Review. Washington, D.C., USDA (United States Department of Agriculture), Agriculture Handbook 480, 137–170.

Batty, R. 1975. Sweet potato. Hortus, 22, 35–42.

Beck, B.D.A. 1959. Annual report. Lagos, Nigeria, Federal Department of Agricultural Research.

1960. Annual report. Lagos, Nigeria, Federal Department of Agricultural Research.

1971. Breeding goals in a cassava breeding programme in West Africa. Paper prepared for the Guinea Zone Conference, University of Ibadan, Nigeria. 5 p.

Beck, B.D.A., and Chant, S.R. 1958. Preliminary investigation on the effect of mosaic virus on *Manihot utilissima* Pohl in Nigeria. Tropical Agriculture (Trinidad), 35(1), 59–64.

Biggs, C.E.J. 1940. Sweet potatoes. In Tothill, J.D., ed., Agriculture in Uganda. London, England, Oxford University Press, 126–128.

Blin, H. 1905. La fumure du manioc. Bulletin économique de Madagascar, 3, 419–421.

Bock, K.R., and Guthrie, E.J. 1976. Recent advances in research on cassava viruses in East Africa. In Nestel, B.L., ed., African Cassava Mosaic: Report of an Interdisciplinary Workshop Held at Muguga, Kenya, 19–22 February 1976. Ottawa, Ont., Canada, IDRC, IDRC-071e, 11–16.

1978. African mosaic disease in Kenya. In Brekelbaum, T., Bellotti, A., and Lozano, J.C., ed., Proceedings: Cassava Protection Workshop, CIAT, Cali, Colombia, 7–12 November, 1977. Cali, Colombia, CIAT, 41–44.

Bock, K.R., Guthrie, E.J., and Meredith, G. 1978. Distribution, host range, properties, and purification of cassava latent virus, a geminivirus. Annals of Applied Biology, 90, 361–367.

Bondar, G. 1912. Una nova molestia bacteriana das hastes da mandioca. Charara e quintaes, 5, 15–18.

Buyckx, E.J.E. 1962. Précis des maladies et des insectes nuisibles rencontrés sur les plantes cultivées au Congo, au Rwanda et au Burundi. Brussels, Belgium, INEAC, Hors série.

Camerman, A., and van Bellinghen, A. 1973. Conclusions des essais de triage et de sondage de population de patates douces à Rubona et à Karama. Rubona, Rwanda, ISAR, Note technique 22.

Chadha, Y.R. 1961. Source of starch in Commonwealth territories. Tropical Science, 3(3), 101–113.

Chant, S.R. 1959. Note on the inactivation of mosaic virus in cassava (*Manihot utilissima* Pohl) by heat treatment. Empire Journal of Experimental Agriculture, 27, 55–58.

Chant, S.R., Bateman, J.G., and Bates, D.C. 1971. Effect of cassava mosaic virus infection on the metabolism of cassava leaves. Tropical Agriculture (Trinidad), 48(3), 263–270.

Chapman, T. 1965. Some investigations into the factors limiting yields of Lisbon yam (*Dioscorea alata* L.) under Trinidad conditions. Tropical Agriculture (Trinidad), 42(2), 145.

Chapola, G.M. 1980. Effect of severity of mosaic on yield of cassava. Paper prepared for the First Triennial Root Crops Symposium of the International Society for Tropical Root Crops — Africa Branch, 8–12 September, Ibadan, Nigeria (unpublished).

CIAT (Centro Internacional de Agricultura Tropical). 1975. 1974 annual report. Cali, Colombia, CIAT.

Clarke, R.T. 1962. Effect of some resting treatments on a tropical soil. Empire Journal of Experimental Agriculture, 30, 57–62.

Cockerham, K.L., Deen, O.T., Christian, M.B., and Newsom, L.D. 1954. Biology of the sweet potato weevil. Louisiana Agriculture Experiment Station Bulletin, 483.

COPR (Centre for Overseas Pest Research). 1977. Control of *Zonocerus variegatus* (L.) in Nigeria. Final report and recommendations. London, England, COPR.

Comstock, R.E., Robinson, H.F., and Harvey, P.H. 1949. Breeding procedure designed to make maximum use of both general and specific combining ability. Agronomy Journal, 41, 360–367.

Cours, G. 1936. Notes sur quelques essais entrepris sur le manioc à la Station agricole de l'Alaotra. Bulletin économique de Madagascar, 6, 162–171.
1951. Le manioc à Madagascar. Agronomie tropicale, 6(1–2), 76–77.
1953. Le manioc. Recherche agronomique de Madagascar, compte rendu, 2, 78–88.
Coursey, D.G. 1967. Yams. London, England, Longmans, Green, and Co. Ltd, Tropical Agriculture Series. 230 p.
Cuthbert, F.P., and Jones, A. 1972. Resistance in sweet potatoes to Coleoptera increased by recurrent selection. Journal of Economic Entomology, 65, 1655–1658.
Dapper, O. 1668. Description de l'Afrique en 1668. Wolfgang trad du Flamand, 314.
Davis, L. 1976. Review of cassava research in Malawi. Lilongwe, Malawi, Chitedze Research Station (Box 158) (unpublished).
DeBach, P. 1974. Biological control by natural enemies. New York, Cambridge University Press. 323 p.
De Bruijn, G.H. 1980. Progress report on the co-operative research project: study of the significance of cassava in the cropping patterns of dry areas of Kenya, with special reference to the Eastern Province. A project in the framework of research co-operation between the Ministry of Agriculture and the University of Nairobi. Nairobi, Kenya, University of Nairobi, Department of Crop Science (internal report).
Delepierre, G. 1974. Les régions agricoles du Rwanda. Rubona, Rwanda, ISAR.
Detilleux, E. 1958. La culture de la patate douce dans la région d'Elisabethville. Bulletin d'informations, INEAC, 7, 237–260.
Doku, E.V. 1967. Root crops in Ghana. Paper prepared for the First Symposium of the International Society for Tropical Root Crops, University of the West Indies, St. Augustine, Trinidad, 2–8 April 1967, 1, 39–65.
Doughty, L.R. 1958. Cassava breeding for resistance to mosaic and brown streak viruses. In EAAFRO, Annual Report. Muguga, Kenya, EAAFRO, 48–55.
Doutt, R.L., and DeBach, P. 1964. Some biological control concepts and questions. In DeBach, P. ed., Biological Control of Insect Pests and Weeds. New York, Reinhold Press, 118–142.
EAAFRO (East African Agriculture and Forestry Research Organization). 1949. Report of the plant pathologist. Muguga, Kenya, EAAFRO, 44.
1951. Annual report. Muguga, Kenya, EAAFRO, 88.
1955. Annual report. Muguga, Kenya, EAAFRO, 36–39.
Eberhart, S.A., Harrison, M.N., and Ogada, F. 1967. Comprehensive breeding system. Züchter, 37, 169–174.
Enyi, B.A.C. 1970. Yams in Africa. In International Society for Tropical Root Crops, Tropical Root and Tuber Crops Tomorrow. Proceedings of the Second International Symposium on Tropical Root and Tuber Crops, Honolulu and Kapaca, Hawaii, August 1970. Honolulu, University of Hawaii, 90–93.
1972. Cassava varietal trial. Tropical Root and Tuber Crops Newsletter, 5, 7–11.
1973. Growth rates of three cassava varieties (*Manihot esculenta* Crantz) under varying population densities. Journal of Agricultural Science, 81(1), 15–28.
Ezumah, H.C. 1980. Quelques recommandations sur les pratiques culturales du manioc au Zaïre. Kinshasa, Zaire, PRONAM, Monographe 1. 13 p.
Ezumah, H.C., Haynes, P.H., and Nwanze, K.F. 1979. Selections for resistance to cassava mealybug (*P. manihoti*) and effects of some cultural practices on incidence and severity in Bas Zaire. Paper prepared for the Fifth International Symposium on Tropical Root and Tuber Crops, Manila, Philippines, 17–21 September.
Ezumah, H.C., and Okigbo, B.N. 1980. Cassava planting systems in Africa. In Weber, E.J., Toro M., J.C., and Graham, M., ed., Cassava Cultural Practices: Proceedings of a Workshop Held in Salvador, Bahía, Brazil, 18–21 March 1980. Ottawa, Canada, IDRC, IDRC-151e, 44–49.
Faan, H.C., Chou, L.K., and Leung, H.M. 1959. Field control of bacterial wilt of sweet potato. Review of Applied Mycology, 39, 497.
Fabres, G., and Boussiengue, J. 1980. Bioécologie de la cochenille du manioc (*Phenacoccus manihoti* Homop. : Pseudococcidae) et de ses populations en République Populaire de Congo. I : Cycle évolutif et paramètres biologiques. Brazzaville, Congo, ORSTOM (Office de la recherche scientifique et technique outre-mer). 16 p.
FAO (Food and Agriculture Organization of the United Nations). 1975. Production yearbook. Rome, Italy, FAO, 29.
1976. Production yearbook. Rome, Italy, FAO, 30.
1977. Production yearbook. Rome, Italy, FAO, 31.
Faulkner, A.C. 1932. Annual report. Lagos, Nigeria, Department of Agriculture, Southern Region.
Ferguson, T.U. 1970. Status and future of yams in the Caribbean. In International Society for Tropical Root Crops, Tropical Root and Tuber Crops Tomorrow. Proceedings of the Second International Symposium on Tropical Root and Tuber Crops, Honolulu and Kapaa, Hawaii, August 1970. Honolulu, University of Hawaii, 2, 28–30.
Ferguson, T.U., and Haynes, P.H. 1970. Response of yams (*Dioscorea* spp.) to nitrogen, phosphorus, potassium and organic fertilisers. In International Society for Tropical Root Crops, Tropical Root and Tuber Crops Tomorrow. Proceedings of the Second International Symposium

on Tropical Root and Tuber Crops, Honolulu and Kapaa, Hawaii, August 1970. Honolulu, University of Hawaii, 1, 93–96.

Fröhlich, G., and Rodewald, W. 1970. Pests and diseases of tropical crops and their control. London, England, Pergamon Press, 177–179.

Gardner, C.O. 1961. Evaluation of effects of mass selection and seed irradiation with thermal neutrons on yield of corn. Crop Science, 1, 241–245.

Gilbert, N. 1967. Additive combining abilities fitted to plant breeding data. Biometrics, 23, 245–249.

Girling, D.J., Bennett, F.D., and Yaseen, M. 1979. Biological control of the green mite *Mononychellus tanajoa* (Bondar) (Acarina: Tetranychidae) in Africa. In Brekelbaum, T., Bellotti, A., and Lozano, J.C., ed., Proceedings: Cassava Protection Workshop, CIAT, Cali, Colombia, 7–12 November, 1977. Cali, Colombia, CIAT, 165–170.

Godfrey-Sam-Aggrey, W. 1980. Annual report: agro-economic studies on cassava in Swaziland. Luyengo, Swaziland, University College of Swaziland, Department of Crop Production.

Godfrey-Sam-Aggrey, W., Haque, I., and Fendru, I. 1978–79. Agro-economic studies on cassava in Swaziland: IFS-sponsored project. Luyengo, Swaziland, University College of Swaziland, Department of Crop Production.

Golding, F.D. 1936. Cassava mosaic in southern Nigeria. Bulletin Agriculture Department (Nigeria), 11, 1.

Groth, B.H.A. 1911. Sweet potato. New York, D. Appleton and Company Agents.

Hahn, S.K. 1975. In IITA, Annual Report, 1974. Ibadan, Nigeria, IITA.

1977a. Sweet potato. In Alvim, R.T., and Kozlowski, T.T., ed., Ecophysiology of Tropical Crops. New York, Academic Press, 237–248.

1977b. In IITA, Annual Report, 1976. Ibadan, Nigeria, IITA.

1979. Effect of viruses (SPVD) on growth and yield of sweet potato. Experimental Agriculture, 15, 253–256.

Hahn, S.K., and Howland, A.K. 1972. Breeding for resistance to cassava mosaic disease. In IITA, Report on an IITA/IDRC-Sponsored Workshop on Cassava Mosaic, Ibadan, December 1972. Ibadan, Nigeria, IITA.

Hahn, S.K., Howland, A.K., and Okoli, C.A. 1974. Breeding for resistance to cassava bacterial blight at IITA. In Okpala, E.U., and Glaser, H.J., ed., Workshop on Cassava Bacterial Blight in Nigeria. Umudike, Nigeria, Federal Agricultural Research and Training Station, 11–14.

Hahn, S.K., Howland, A.K., and Terry, E.R. 1980. Correlated resistance of cassava to mosaic and bacterial blight diseases. Euphytica, 29(2), 305–311.

Hahn, S.K., and Hozyo, Y. 1980. Sweet potato and yams. Paper prepared for the Symposium on Potential Productivity of Field Crops under Different Environments, International Rice Research Institute, Los Baños, Philippines. 38 p.

Hahn, S.K., Terry, E.R., and Leuschner, K. 1980. Breeding cassava for resistance to cassava mosaic disease. Euphytica, 29.

1981. Resistance of sweet potato to virus complex. HortScience, 16(4), 535–537.

Hahn, S.K., Terry, E.R., Leuschner, K., Akobundu, I.O., Okali, C., and Lal, R. 1979. Cassava improvement in Africa. Field Crops Research, 2, 193–226.

Hahn, S.K., and Williams, L. 1973. Investigations on cassava in the Republic of Zaire. Report to the Commissaire d'état à l'Agriculture, Republic of Zaire, 12–30 March. Kinshasa, Republic of Zaire, Commissaire d'état à l'Agriculture. 12 p.

Haque, I., Godfrey-Sam-Aggrey, W., and Nzima, D.M.S. 1979. Root crop research in Swaziland. Luyengo, Swaziland, University College of Swaziland, Department of Crop Production.

Harlan, H.V., and Martini, M.L. 1938. Effect of natural selection in a mixture of barley varieties. Journal of Agricultural Research, 57, 189–199.

Hawkins, R. 1593. Observations of Sir Richard Hawkins, Knt. in his voyage into the South Seas in the year 1593. In Bethume, C.R.D., ed., Hakluyt Society, Account XXVII. London, England, Hakluyt Society (reprinted from the 1622 edition).

Heys, n.d. Report to the German agency for technical co-operation Ltd (GTZ) on a study to determine the feasibility of an extension of cassava production on a commercial scale in Malawi. Ibadan, Nigeria, IITA (unpublished).

Hollings, M. 1965. Disease control through virus-free stock. American Review of Phytopathology, 3, 367–396.

Hollings, M., Stone, O.M., and Bock, K.R. 1976. Purification and properties of sweet potato mild mottle, a whitefly borne virus from sweet potato (*Ipomoea batatas*) in East Africa. Annals of Applied Biology, 82, 511–528.

Howell, M.J., and Wittwer, S.H. 1955. Further studies on the effects of 2,4-D on flowering in the sweet potato. Proceedings of the American Society for Horticultural Science, 66, 279–283.

Hussey, N.W., and Scopes, N.E.A. 1977. Introduction of natural enemies for pest control in glasshouses: ecological consideration. In Ridgway, R.L., and Vinson, S.B., ed., Biological Control by Augmentation of Natural Enemies. Insect and Mite Control with Parasites and Predators. New York, Plenum Press, 349–377.

IFAN (Institut fondamental d'Afrique Noire). 1951. Première conférence, Internationale des Africanistes de l'ouest, Comptes Rendus. Dakar, Sénégal, IFAN.

IITA (International Institute of Tropical Agriculture). 1972. Annual report of the Root and Tuber Improvement Programme. Ibadan, Nigeria, IITA.

1973. Annual report, 1972. Ibadan, Nigeria, IITA.
1974. Annual report, 1973. Ibadan, Nigeria, IITA.
1975. Annual report, 1974. Ibadan, Nigeria, IITA.
1976. Annual report, 1975. Ibadan, Nigeria, IITA.
1977. Annual report, 1976. Ibadan, Nigeria, IITA.
1978. Annual report, 1977. Ibadan, Nigeria, IITA.
1979. Annual report, 1978. Ibadan, Nigeria, IITA.

Irving, H. 1956. Fertiliser experiments with yams in eastern Nigeria 1949–1951. Tropical Agriculture (Trinidad), 33(1), 67.

ISAR (Institut des sciences agronomiques du Rwanda). 1964. Annual report. Rubona, Rwanda, ISAR.
1965. Annual report. Rubona, Rwanda, ISAR.
1966. Annual report. Rubona, Rwanda, ISAR.
1967. Annual report. Rubona, Rwanda, ISAR.
1968. Annual report. Rubona, Rwanda, ISAR.
1969. Annual report. Rubona, Rwanda, ISAR.

Jacoby, T. 1967. Nutrition and manuring of tropical root crops. Hannover, Germany, Verkaufsgemeinschaft Deutscher Kaliwerke, GMBH.

Janssens, M.J.J. 1980. Detection of non-additive gene action with nested polycross designs. I: Procedure assuming epistasis negligible. Silvae Genetica, 29(3–4), 138–140.

Jennings, D.L. 1957. Further studies in breeding cassava for virus resistance. East African Agricultural Journal, 22(4), 213–219.
1970. Cassava in Africa. Field Crop Abstracts, 23(3), 271–275.
1972. Breeding for resistance to cassava viruses in East Africa. In IITA, Report on an IITA/IDRC-Sponsored Workshop on Cassava Mosaic, Ibadan, December 1972. Ibadan, Nigeria, IITA.
1976. Breeding for resistance to African cassava mosaic disease: progress and prospects. In Nestel, B.L., ed., African Cassava Mosaic: Report of an Interdisciplinary Workshop Held at Muguga, Kenya, 19–22 February 1976. Ottawa, Canada, IDRC, IDRC-071e, 39–44.

Jones, A. 1965. Proposed breeding procedure for sweet potato. Crop Science, 5, 191–192.
1970. Phenotypic, genotypic, and environmental correlations in sweet potatoes. Journal of the American Society for Horticultural Science, 95, 326–330.
1972. Sweet potato population changes during six random crossing generations. Journal of the American Society for Horticultural Science, 97, 102–107.
1977. Pesticides increase true seed production of sweet potato. HortScience, 12, 165–167.

Jones, W.O. 1957. Manioc: an example of innovation in African economies. Economic Development and Cultural Change, 5(2), 97–117.
1959. Manioc in Africa. Palo Alto, California, USA, Stanford University Press.

Kartha, A.K., and Gamborg, D.L. 1975. Elimination of cassava mosaic disease by meristem culture. Phytopathology, 65, 826–828.

Kenya, Central Bureau of Statistics. 1978. Statistical abstracts, Republic of Kenya. Nairobi, Kenya, Central Bureau of Statistics, Ministry of Economic Planning and Community Affairs.

Kenya, Government of. 1979. Development plan 1979–1983. Nairobi, Kenya, Government of Kenya, 222.

Knavel, D.E., and Lasheen, A.M. 1969. Association of flowering with nutrition in the sweet potato, *Ipomoea batatas* L. Journal of the American Society for Horticultural Science, 94, 675–677.

Koli, S.E. 1973. Response of yam (*Dioscorea rotundata*) to fertiliser application in northern Ghana. Journal of Agricultural Science, 80(2), 245–249.

Konishi, M., and Ito, Y. 1973. Early entomology in East Asia. In Smith, R.F., et al., ed., History of Entomology, California, USA, Annual Review Inc., 1–20.

Lacerda. 1798. Lacerda's journey to Cazembe in 1798. (Translated and annotated by R.F. Burton, Royal Geographical Society, London, 1873.)

Lacourbe, de. 1635. Premier voyage du sieur de la Courbe fait à la cote d'Afrique en 1635. Paris, France, P. Cultru, Champion et Larose, 1913.

Lefevre, P. 1935. Quelques considérations sur le mosaïque du manioc. Bulletin agricole du Congo Belge, 26, 442–447.

Lefevre, P.C. 1948. *Acraea acerata* Hew. parasite de la patate douce. Bulletin agricole du Congo Belge, 39, 49–76.

Lemarchand, G. 1956. La sélection de la patate douce à Mulungu. Bulletin d'informations, INEAC, 5, 237–254.

Leuschner, K. 1976. In IITA, Annual Report, 1975. Ibadan, Nigeria, IITA.
1977a. In IITA, Annual Report, 1976. Ibadan, Nigeria, IITA.
1977b. Preliminary observations on mealybug (Hemiptera, Pseudococcidae) in Zaire and a projected outline for subsequent work. In Nwanze, K.F., and Leuschner, K., ed., Proceedings of the International Workshop on the Cassava Mealybug *Phenacoccus manihoti* Mat.-Ferr. (Pseudococcidae). M'vuazi, Zaire, INERA, Proceedings Series 1, 15–19.

Lozano, J.C. 1975. Bacterial blight of cassava. PANS, 21, 38–43.
1977. Cassava. In Hewitt, W.B., and Chiarappa, L., ed., Plant Health and Quarantine in International Transfer of Genetic Resources. Ohio, USA, CRC Press. 346 p.

Lozano, J.C., and Sequeira, L. 1974. Bacterial blight of cassava in Colombia. I: etiology. Phytopathology, 64, 74–82.

Lozano, J.C., and Terry, E.R. 1976. Cassava diseases and their control. Noticias Fitopatológicas, 5, 38–44.
MacDonald, A.S. 1963. Sweet potatoes, with particular reference to the tropics. Field Crop Abstracts, 16(4), 191–225.
Mahungu, N. 1979. Relationship between fresh tuberous root yield and its components in sweet potato, *Ipomoea batatas* L. Lam. Ibadan, Nigeria, University of Ibadan (Master's thesis).
Mann, W.S. 1963. Fertiliser use in eastern Nigeria. Enugu, Nigeria, Government Printer, Technical Bulletin 5.
Maraite, H., and Perreaux, D. 1978. Comparative symptom development in cassava after infection by *Xanthomonas manihotis* or *X. cassavae* under controlled conditions. In Terry, E.R., Persley, G., and Cook, S.C.A., ed., Cassava Bacterial Blight in Africa: Past, Present, and Future. Report of an Interdisciplinary Workshop Held at IITA, Ibadan, Nigeria, 26–30 June 1978. London, England, COPR.
Martin, W.J., and Dukes, P.D. 1975. Bacterial soft rot of sweet potato plants. Proceedings of the American Phytopathological Society, 2, 156.
Matile-Ferrero, D. 1977. Cochenille nouvelle nuisible au manioc en Afrique equatoriale, *Phenacoccus manihoti* n. sp. (Homoptera, Coccidoidae, Pseudococcidae). Annales de la société entomologique de France, 13, 145–151.
Mauritius, Sugar Industry Research Institute. 1971. Annual report. Réduit, Mauritius, Sugar Industry Research Institute.
1972. Annual report. Réduit, Mauritius, Sugar Industry Research Institute.
1973. Annual report. Réduit, Mauritius, Sugar Industry Research Institute.
1974. Annual report. Réduit, Mauritius, Sugar Industry Research Institute.
Miles, A.C. 1934. Report of the Department of Agriculture, 1934–35. Accra, Ghana, Department of Agriculture.
Moran, E.F. 1976. Manioc deserves more recognition in tropical farming. World Crops, 28(4), 184–188.
Mukiibi, J. 1977. Effect of mosaic on yield of sweet potatoes in Uganda. In Cock, J., MacIntyre, R., and Graham, M., ed., Proceedings of the Fourth Symposium of the International Society for Tropical Root Crops Held at CIAT, Cali, Colombia, 1–7 August 1976. Ottawa, Canada, IDRC, IDRC-080e, 169–170.
Murata, Y., Kumara, A., and Ishiyee, L. 1976. Photosynthesis and ecology of crop plants. Tokyo, Japan, Cultural Association of Agriculture, Forestry, and Fishery, 204–233.
Murdoch, D. 1970. Soils and land capability of Swaziland. Mbabane, Swaziland, Ministry of Agriculture.
Nakagawa, Y. 1970. Yams (*Dioscorea* sp.) as a food crop in some Pacific islands. In International Society for Tropical Root Crops, Tropical Root and Tuber Crops Tomorrow. Proceedings of the Second International Symposium on Tropical Root and Tuber Crops, Honolulu and Kapaa, Hawaii, August 1970. Honolulu, University of Hawaii, 2, 23–24.
Nichols, R.F.W. 1947. Breeding cassava for virus resistance. East African Agricultural Journal, 12, 184–194.
1950. Brown streak disease of cassava. Distribution, climatic effects and diagnostic symptoms. East African Agriculture and Forestry Journal, 15, 154–160.
Nielsen, L.W., and Terry, E.R. 1977. Sweet potato. In Hewitt, W.B., and Chiarappa, L., ed., Plant Health and Quarantine in International Transfer of Genetic Resources. Ohio, USA, CRC Press. 346 p.
Nwanze, K.F. 1978. Biology of the cassava mealybug, *Phenacoccus manihoti* Mat.-Ferr., in the Republic of Zaire. In Nwanze, K.F., and Leuschner, K., ed., Proceedings of the International Workshop on the Cassava Mealybug *Phenacoccus manihoti* Mat.-Ferr. (Pseudococcidae). M'vuazi, Zaire, INERA (Institut national pour l'étude et la recherche agronomique), Proceedings Series 1, 20–28.
Nye, P.H. 1954. Fertiliser responses in the Gold Coast in relation to time and method of application. Empire Journal of Experimental Agriculture, 22(86), 101.
Nyiira, Z.M. 1972. Report of investigation on cassava mite, *Mononychellus tanajoa* Bondar. Kampala, Uganda, Kawanda Research Station. 14 p.
1976. Advances in research on the economic significance of the cassava green mite (*Mononychellus tanajoa*) in Uganda. In Terry, E.R., and MacIntyre, R., ed., International Exchange and Testing of Cassava Germ Plasm in Africa: Proceedings of an Interdisciplinary Workshop Held at IITA, Ibadan, Nigeria, 17–21 November 1975. Ottawa, Canada, IDRC, IDRC-063e, 27–29.
1977. Population dynamics of the green cassava mite and its predator *Oligota*. In Cock, J., MacIntyre, R., and Graham, M., ed., Proceedings of the Fourth Symposium of the International Society for Tropical Root Crops Held at CIAT, Cali, Colombia, 1–7 August 1976. Ottawa, IDRC, IDRC-080e, 193–197.
1979. *Mononychellus tanajoa* (Bondar): biology, ecology and economic importance. In Brekelbaum, T., Bellotti, A., and Lozano, J.C., ed., Proceedings: Cassava Protection Workshop, CIAT, Cali, Colombia, 7–12 November, 1977. Cali, Colombia, CIAT, 155–159.
Oatman, E.R., and McMurtry, J.A. 1966. Biological control of the two-spotted mite on strawberry in southern California. Journal of Economic Entomology, 59, 422–439.
Obi, J.K. 1959. Standard D. N.P.K. experiments. Samaru, Nigeria, Regional Research Station, Technical Report 8.
Obigbesan, G.O., and Matuluko, E.O. 1977. Effect

of potassium and bacterial blight on the yield and chemical composition of cassava cultivars. In Cock, J., MacIntyre, R., and Graham, M., ed., Proceedings of the Fourth Symposium of the International Society for Tropical Root Crops Held at CIAT, Cali, Colombia, 1–7 August 1976. Ottawa, Canada, IDRC, IDRC-080e, 185–188.

Onyango, D.M., and Ramos, A.H. 1978. Bacterial blight of cassava in Kenya. In Terry, E.R., Persley, G.J., and Cook, S.C.A., ed., Cassava Bacterial Blight in Africa: Past, Present, and Future. Report of an Interdisciplinary Workshop Held at IITA, Ibadan, Nigeria, 26–30 June 1978. London, England, COPR, 25–28.

PANS. 1978 Pest control in tropical root crops. London, England, COPR, PANS Manual, 4. 235 p.

Parlevliet, J.E., and Zadocks, J.C. 1977. Integrated concept of disease resistance: a new view including horizontal and vertical resistance in plants. Euphytica, 26, 5–21.

Perreaux, D., Terry, E.R., and Persley, G.J. 1978. Studies on screening methods of cassava resistance to bacterial blight. In Terry, E.R., Persley, G.J., and Cook, S.C.A., ed., Cassava Bacterial Blight in Africa: Past, Present, and Future. Report of an Interdisciplinary Workshop Held at IITA, Ibadan, Nigeria, 26–30 June 1978. London, England, COPR.

Persley, G.J. 1979. Studies on the survival and transmission of *Xanthomonas manihotis* on cassava seed. Annals of Applied Biology, 93, 159–163.

Prior, A.J. 1975. Should cassava be fertilized or not? Farming in Zanzibar, 9(2), 3–4.

PRONAM (Programme national manioc). 1975. Rapport annuel. Kinshasa, Zaire, PRONAM.

1976. Rapport annuel. Kinshasa, Zaire, PRONAM.

1977. Rapport annuel. Kinshasa, Zaire, PRONAM.

1978. Rapport annuel. Kinshasa, Zaire, PRONAM.

1979. Rapport annuel. Kinshasa, Zaire, PRONAM.

Pynaert, L. 1928. Le manioc. Bulletin agricole du Congo Belge, 19, 163.

Roca, W.M., Rodriguez, A., Patena, L.F., Barba, R.C., and Toro M., J.C. 1980. Improvement of a propagation technique for cassava using single leaf bud cuttings. Cali, Colombia, CIAT, Cassava Newsletter, 8, 4–5.

Rollet, C. 1926. Le manioc à Madagascar. Revue internationale de botanique appliquée et d'agriculture tropicale, 6(55), 52–159.

Russell, L.M. 1957. Synonyms of *Bemisia tabaci* (Grennadius). Bulletin of the Brooklyn Entomological Society, 52(5), 122–123.

Salifu, E.B. 1980. ADMARC producer prices for 1979/80 planted crops: effective 1st April, 1980. Southern Adviser, March–April, 13(3–4).

Samways, M. 1979. Alcohol from cassava in Brazil: pest threatens yields. World Crops, 31(5), 181–186.

Scaife, A. 1966. Cassava in western Tanzania. Tanzania, Western Research Centre Progress Report, 3. 2 p.

1968. Effect of a cassava "fallow" and various manurial treatments on cotton at Ukiriguru, Tanzania. East African Agriculture and Forestry Journal, 33(3), 231–235.

Schaefers, G.A., and Terry, E.R. 1976. Insect transmission of sweet potato disease agents in Nigeria. Phytopathology, 47, 582–590.

Schutz, W.M. et al. 1968. Intergenotypic competition in plant populations. Crop Science, 8, 61–66.

Shaw, S.M., and Oogarah, P. 1972. Report of field work carried out by the Agriculture Mission of the Republic of China to Mauritius (covering the period from Feb. to Aug., 1971, series 3). Revue agricole et sucrière de l'île Maurice, 51(2), 97–114.

Sheffield, F.M.L. 1953. Viruses of sweet potato in parts of Africa. Empire Journal of Experimental Agriculture, 21, 184–189.

1957. Virus diseases of sweet potato in East Africa. I: identification of the viruses and their insect vectors. Phytopathology, 47, 582–590.

1962. In Buyckx, E.J.E., Précis des maladies et des insectes nuisibles rencontré sur les plantes cultivées au Congo, au Rwanda et au Burundi. Brussels, Belgium, INEAC, Hors série.

Shukla, P.T. 1975. Cassava variety catalogue. Mwanza, Tanzania, Ukiriguru.

Singh, T.P. 1979. Sélection du manioc pour la résistance aux pestes et maladies en République du Zaire. Kinshasa, Zaire, PRONAM, Série 3. 16 p.

Stephens, D. 1956. Fertiliser trials 1948–1951. New Gold Coast Farmer, 1(1), 9.

Storey, H.H. 1930. Second annual report of the East African Agricultural Research Station. Amani, Tanzania, East African Agricultural Research Station.

1935. Report of the plant pathologist. Amani, Tanzania, East African Agricultural Research Station, 12–16.

1936. Virus diseases of East African plants. VI: progress report on studies of the diseases of cassava. East African Agricultural Journal, 2, 34–39.

Storey, H.H., and Nichols, R.F.W. 1938. Studies on the mosaic diseases of cassava. Annals of Applied Biology, 25(4), 790–806.

Stuhlman, F. 1909. Beitrage zur Kulturgeschichte von Ostafrika Bd.X of Deutsch-Ost-Afrika, Berlin.

Suge, H. 1977. Promotion of flowering in sweet potato by gibberellin A_3 and A_7. Japanese Journal of Breeding, 27, 251–256.

Suneson, C.A. 1956. Evolutionary plant breeding method. Agronomy Journal, 48, 188–190.

Swaziland, Central Statistical Office. 1976. Annual statistical bulletin. Mbabane, Swaziland, Central Statistical Office.

1977. Annual survey of Swazi nation land. Mbabane, Swaziland, Central Statistical Office.

Terry, E.R. 1978a. In IITA, Annual report, 1977. Ibadan, Nigeria, IITA, 53.

1978b. Cassava bacterial blight in Africa. In Terry, E.R., Persley, G.J., and Cook, S.C.A., ed., Cassava Bacterial Blight in Africa: Past, Present, and Future. Report of an Interdisciplinary Workshop Held at IITA, Ibadan, Nigeria, 26–30 June 1978. London, England, COPR.

1978c. Cassava bacterial diseases. In Brekelbaum, T., Belotti, A., and Lozano, J.C., ed., Proceedings: Cassava Protection Workshop, CIAT, Cali, Colombia, 7–12 November, 1977. Cali, Colombia, CIAT, 75–84.

1978d. Disease problems in root crops production. In Ene, L.S.O., Okereke, H.E., Odurukwe, S.O., Okoli, O.O., and Arene, O.B., ed., Proceedings of the First National Seminar on Root and Tuber Crops. Umudike, Nigeria, NRCRI (National Root Crops Research Institute).

1979. Screening sweet potato for resistance to viruses. Paper prepared for the 1979 IITA Research Conference, IITA, Ibadan, Nigeria, October. 18 p. (mimeographed document).

Terry, E.R., and Hahn, S.K. 1980. Effect of cassava mosaic disease on growth and yield of a local and an improved variety of cassava. Tropical Pest Management, March, 26(1), 34–37.

Thompson, N.R., Wurster, R.T., and Sayre, K.D. 1977. Utilisation of the potato, *Solanum tuberosum* in the tropics. Lima, Peru, CIP (Centro Internacional de la Papa).

Tuttle, D.M., Baker, E.W., and Sales, F.M. 1977. Report, Agricultural Experimental Station. Tucson, Arizona, USA, University of Arizona, 2, 1–6.

Uganda, Ministry of Planning and Economic Development. 1979. Agricultural statistics. Kampala, Uganda, Ministry of Planning and Economic Development.

Umanah, E.E. 1973. Effects of different rates of N.P.K. fertilisers on yield and storage properties of white yam (*Dioscorea rotundata*). Paper prepared for the Third International Symposium on Tropical Root Crops, IITA, Ibadan, Nigeria, 2–9 December 1973. Ibadan, Nigeria, IITA.

Uriyo, A.P., and Kesseba, A. 1973. Response of sweet potato (*Ipomoea batatas* Poir) to phosphorus and potassium fertilizers on a red oxisol soil in Tanzania. Beitrage zur Tropischen und Subtropischen Landwirtschaft und Tropenveterinaermedizin, 11(3), 299–302.

Vavilov, N.I. 1949. Chronica Botanica, 13, 43.

Vine, H. 1953. Experiments on the maintenance of soil fertility at Ibadan, Nigeria. Empire Journal of Experimental Agriculture, 21(81), 65.

Wiehe, P.O., and Dowson, W.J. 1953. Bacterial disease of cassava (*Manihot utilissima*) in Nyasaland. Empire Journal of Experimental Agriculture, 21(82), 141–143.

Wood, R.C. 1933. Experiments with yams in Trinidad 1931–1933. Empire Journal of Experimental Agriculture, 1(14), 316.

Wright, W.A. 1961. Potato industry of Mauritius. Revue agricole et sucrière de l'ile maurice, 40(3), 138–147.